WORKSHEETS
FOR CLASSROOM OR LAB PRACTICE
EDUMEDIA SERVICES, INC.

BEGINNING AND INTERMEDIATE ALGEBRA
THIRD EDITION

John Tobey
North Shore Community College

Jeffrey Slater
North Shore Community College

Jamie Blair
Orange Coast College

Prentice Hall
is an imprint of

The author and publisher of this book have used their best efforts in preparing this book. These efforts include the development, research, and testing of the theories and programs to determine their effectiveness. The author and publisher make no warranty of any kind, expresses or implied, with regard to these programs or the documentation contained in this book. The author and publisher shall not be liable in any event for incidental or consequential damages in connection with, or arising out of, the furnishing, performance, or use of these programs.

Reproduced by Pearson Prentice Hall from electronic files supplied by the author.

Copyright ©2010 Pearson Education, Inc.
Publishing as Pearson Prentice Hall, Upper Saddle River, NJ 07458.

All rights reserved. No part of this publication may be reproduced, stored in a retrieval system, or transmitted, in any form or by any means, electronic, mechanical, photocopying, recording, or otherwise, without the prior written permission of the publisher. Printed in the United States of America.

ISBN-13: 978-0-321-58585-1
ISBN-10: 0-321-58585-2

1 2 3 4 5 6 OPM 12 11 10 09

Prentice Hall
is an imprint of

www.pearsonhighered.com

Worksheets

Beginning and Intermediate Algebra, Third Edition

Table of Contents

Chapter 0   **1**
- Section 0.1   1
- Section 0.2   3
- Section 0.3   5
- Section 0.4   7
- Section 0.5   9

Chapter 1   **11**
- Section 1.1   11
- Section 1.2   13
- Section 1.3   15
- Section 1.4   17
- Section 1.5   19
- Section 1.6   21
- Section 1.7   23
- Section 1.8   25
- Section 1.9   27

Chapter 2   **29**
- Section 2.1   29
- Section 2.2   31
- Section 2.3   33
- Section 2.4   35
- Section 2.5   37
- Section 2.6   39
- Section 2.7   41
- Section 2.8   43

Chapter 3   **45**
- Section 3.1   45
- Section 3.2   47
- Section 3.3   49
- Section 3.4   51
- Section 3.5   53
- Section 3.6   55

Chapter 4   **57**
- Section 4.1   57
- Section 4.2   59
- Section 4.3   61
- Section 4.4   63

Chapter 5   **65**
- Section 5.1   65
- Section 5.2   67
- Section 5.3   69
- Section 5.4   71
- Section 5.5   73
- Section 5.6   75

Chapter 6	77		**Chapter 10**	131
Section 6.1	77		Section 10.1	131
Section 6.2	79		Section 10.2	133
Section 6.3	81		Section 10.3	135
Section 6.4	83		Section 10.4	137
Section 6.5	85		Section 10.5	139
Section 6.6	87			
Section 6.7	89			
Chapter 7	91		**Chapter 11**	141
Section 7.1	91		Section 11.1	141
Section 7.2	93		Section 11.2	143
Section 7.3	95		Section 11.3	145
Section 7.4	97		Section 11.4	147
Section 7.5	99			
Section 7.6	101			
Chapter 8	103		**Chapter 12**	149
Section 8.1	103		Section 12.1	149
Section 8.2	105		Section 12.2	151
Section 8.3	107		Section 12.3	153
Section 8.4	109		Section 12.4	155
Section 8.5	111		Section 12.5	157
Section 8.6	113			
Section 8.7	115			
Chapter 9	117		**Answers**	159
Section 9.1	117			
Section 9.2	119			
Section 9.3	121			
Section 9.4	123			
Section 9.5	125			
Section 9.6	127			
Section 9.7	129			

Name _____   Date _____

Practice Set 0.1
Simplifying Fractions

Simplify each fraction.

1. $\dfrac{6}{9}$ 1. _____

2. $\dfrac{12}{24}$ 2. _____

3. $\dfrac{20}{28}$ 3. _____

4. $\dfrac{60}{15}$ 4. _____

Change to a mixed number.

5. $\dfrac{14}{3}$ 5. _____

6. $\dfrac{42}{5}$ 6. _____

7. $\dfrac{54}{4}$ 7. _____

8. $\dfrac{556}{10}$ 8. _____

Name _____   Date _____

Change to an improper fraction.

9. $2\dfrac{1}{3}$

9. _____

10. $3\dfrac{4}{5}$

10. _____

11. $1\dfrac{12}{17}$

11. _____

12. $10\dfrac{7}{10}$

12. _____

Find the missing numerator.

13. $\dfrac{2}{3} = \dfrac{?}{21}$

13. _____

14. $\dfrac{6}{7} = \dfrac{?}{56}$

14. _____

15. $\dfrac{8}{11} = \dfrac{?}{132}$

15. _____

16. $\dfrac{8}{15} = \dfrac{?}{120}$

16. _____

Name _____   Date _____

Practice Set 0.2
Adding and Subtracting Fractions

Find the LCD (least common denominator) of each pair of fractions. Do not combine the fractions; only find the LCD.

1. $\dfrac{2}{3}$ and $\dfrac{5}{9}$ 1. _____

2. $\dfrac{3}{4}$ and $\dfrac{9}{14}$ 2. _____

3. $\dfrac{1}{6}$ and $\dfrac{3}{20}$ 3. _____

4. $\dfrac{13}{18}$ and $\dfrac{23}{30}$ 4. _____

5. $\dfrac{6}{25}$ and $\dfrac{99}{100}$ 5. _____

Combine. Be sure to simplify your answer whenever possible.

6. $\dfrac{1}{4}+\dfrac{1}{3}$ 6. _____

7. $\dfrac{3}{5}+\dfrac{1}{10}$ 7. _____

8. $\dfrac{7}{8}-\dfrac{5}{12}$ 8. _____

Name _____ Date _____

9. $\dfrac{31}{45} - \dfrac{2}{15}$

9. _____

10. $8\dfrac{4}{5} + 3\dfrac{9}{10}$

10. _____

11. $6\dfrac{6}{7} - 3\dfrac{3}{14}$

11. _____

12. $4 - 2\dfrac{1}{2}$

12. _____

13. $\dfrac{7}{9} + 2\dfrac{5}{12}$

13. _____

14. $\dfrac{1}{3} + \dfrac{3}{4} + \dfrac{7}{12}$

14. _____

15. $6\dfrac{3}{11} - 2\dfrac{17}{22}$

15. _____

Name _____   Date _____

Practice Set 0.3
Multiplying and Dividing Fractions

Multiply. Simplify your answer whenever possible.

1. $\dfrac{2}{3} \times \dfrac{6}{7}$ 1. _____

2. $\dfrac{8}{5} \times \dfrac{15}{24}$ 2. _____

3. $\dfrac{36}{55} \times \dfrac{5}{6}$ 3. _____

4. $7\dfrac{1}{2} \times 2\dfrac{2}{5}$ 4. _____

5. $\dfrac{3}{7} \times \dfrac{5}{12} \times \dfrac{8}{35}$ 5. _____

Divide. Simplify your answer whenever possible.

6. $\dfrac{2}{3} \div \dfrac{4}{9}$ 6. _____

7. $\dfrac{15}{7} \div \dfrac{15}{4}$ 7. _____

8. $\dfrac{\frac{6}{7}}{\frac{16}{21}}$ 8. _____

Name _____   Date _____

9. $2\dfrac{4}{9} \div 3\dfrac{2}{3}$

9. _____

10. $6\dfrac{2}{5} \div 2\dfrac{4}{5}$

10. _____

11. $\dfrac{4\dfrac{1}{8}}{\dfrac{3}{4}}$

11. _____

Perform the proper calculations. Reduce your answer whenever possible.

12. Rose noticed that the length of her dining room is $\dfrac{2}{3}$ of the length of her living room, and that the width of her dining room is $\dfrac{4}{5}$ of the width of her living room. What fraction expresses the area of the dining room in relation to the living room?

12. _____

13. A favorite at Bill's Diner is the Big Burger, which is a $\dfrac{3}{4}$-pound hamburger with all the toppings you like. If the kitchen currently has 99 pounds of hamburger in stock, how many Big Burgers can they make?

13. _____

14. Arnold worked a $7\dfrac{1}{2}$ hour shift at work. He spent $\dfrac{3}{5}$ of this time organizing the stock room. How many hours did this task take?

14. _____

15. For an art project, Mario cuts $59\dfrac{1}{2}$ inches of string into pieces that are exactly $4\dfrac{1}{4}$ inches long. How many pieces of string does Mario cut?

15. _____

Name _____   Date _____

Practice Set 0.4
Using Decimals

Write each fraction as a decimal. Write each value in words.

1. $5\frac{3}{5}$ 1. _____

2. $12\frac{5}{8}$ 2. _____

Write each decimal as a fraction in simplified form. Write each value in words.

3. 14.0027   3. _____

4. 32.082   4. _____

Add or subtract.

5. 36.237 + 45.304   5. _____

6. 6.0089 + 42.396   6. _____

7. 14.03 − 7.8932   7. _____

Name _____   Date _____

8. 625.11 − 81.709   8. _____

Multiply or divide.

9. 0.25 × 0.003   9. _____

10. 4.003 × 1.03   10. _____

11. 45.705 ÷ 0.05   11. _____

12. 1.1592 ÷ 0.06   12. _____

Multiply or divide by moving the decimal point.

13. 16.785 × 100   13. _____

14. 0.00703 × 1000   14. _____

15. 59.056 ÷ 1000   15. _____

16. 111.39 ÷ 100,000   16. _____

Name _____ Date _____

Practice Set 0.5
Percents, Rounding, and Estimating

Change to a percent.

1. 0.32

1. _____

2. 0.71

2. _____

3. 0.398

3. _____

4. 3.45

4. _____

5. 0.00576

5. _____

Change to a decimal.

6. 5%

6. _____

7. 7.93%

7. _____

8. 0.08%

8. _____

Name _____ Date _____

9. 623.8%   9. _____

10. 100%   10. _____

Find the following.

11. What is 25% of 200?   11. _____

12. What percent of 50 is 22?   12. _____

13. 65 is what percent of 25?   13. _____

Follow the principles of estimation to find an approximate value. Round each number so that there is one nonzero digit. Do not find the exact value.

14. 722×301 14. _____

15. $287 + 198 + 412 + 159 + 279$ 15. _____

16. $21 \overline{)614{,}320}$ 16. _____

Name _____ Date _____

Practice Set 1.1
Adding Real Numbers

Using the numbers 1–5, identify the following as a (1) whole number, (2) rational number, (3) irrational number, (4) integer, or (5) real number. Remember, a number can be identified as more than one type.

Example: –3? Answer: 2, 4, 5

1. 43

2. –3.2

3. Π

4. $-\dfrac{7}{8}$

5. Find the additive inverse (opposite) for each number in Exercises 1 through 4 above.

6. Find the absolute value for each number in Exercises 1 through 4 above.

Use a real number to represent each situation.

7. You owe your best friend $25.

8. The Johnson family drove 342.7 kilometers yesterday.

1. _____
2. _____
3. _____
4. _____
5. _____
6. _____
7. _____
8. _____

Name _____ Date _____

9. Abagail is $3\frac{1}{2}$ years younger than her brother

9. _____

Add.

10. $-14 + (-23)$

10. _____

11. $-9 + (-4) + 8$

11. _____

12. $57 + (-32) + 90 + (-100)$

12. _____

13. $12.93 + (-3.79)$

13. _____

14. $\frac{2}{9} + \left(-\frac{5}{9}\right)$

14. _____

15. $\frac{5}{12} + \left(-\frac{11}{18}\right)$

15. _____

16. $-\frac{2}{3} + 1\frac{4}{9}$

16. _____

Name _____   Date _____

Practice Set 1.2
Subtracting Real Numbers

Subtract by adding the opposite.

1. $6 - 19$

2. $27 - 47$

3. $-42 - (-53)$

4. $(-201) - 32$

5. $-123 - (-123)$

6. $(-5.6) - 9.03$

7. $(-80.09) - (-76.8)$

Subtract by adding the opposite.

8. $\dfrac{2}{7} - \dfrac{18}{21}$

9. $10 - \left(-\dfrac{3}{5}\right)$

1. _____
2. _____
3. _____
4. _____
5. _____
6. _____
7. _____
8. _____
9. _____

Name _____   Date _____

10. $-\dfrac{11}{15} - \left(-\dfrac{1}{2}\right)$ 10. _____

Change each subtraction operation to "adding the opposite." Then combine the numbers.

11. 6 – (–9) – (–10) – 13   11. _____

12. –30 + 12 – (–15) – 8   12. _____

13. 7.5 – (–1.2) + (–8.02)   13. _____

14. Tinnie overdrew her checking account by $25.32. Quickly, she   14. _____
    ran to the bank and deposited her paycheck of $214.78. Write an
    expression to represent this situation. What is the balance of her
    checking after her deposit?

15. On Groundhog Day, the low temperature in Bismarck, North   15. _____
    Dakota was –15 degrees Fahrenheit and the high temperature in
    Las Vegas, Nevada, was 89 degrees Fahrenheit. Write an
    expression to represent this difference in temperature. How many
    degrees warmer was it in Las Vegas than in Bismarck?

16. In the morning, Maxine began hiking uphill from a valley that is   16. _____
    22 feet below sea level. She ate her lunch on top of a hill whose
    altitude is 179 feet above sea level. Write an expression to
    represent the difference in altitude. How many feet did Maxine
    climb during her hike?

Name _____   Date _____

Practice Set 1.3
Multiplying and Dividing Real Numbers

Multiply. Be sure to write your answer in the simplest form.

1. (16)(14)
   1. _____

2. (–81)(–6)
   2. _____

3. (–120)(–13)
   3. _____

4. $\left(-\dfrac{8}{15}\right)\left(\dfrac{3}{4}\right)$
   4. _____

5. (15.8)(–29.3)
   5. _____

Multiply. You may want to determine the sign of the product before you multiply.

6. (–5)(6)(–3)(–10)
   6. _____

7. (–7)(–2)(–5)(–1)
   7. _____

Name _____   Date _____

8. (–23)(35)(0)(–4)(–81)   8. _____

9. $5(-3)(-2)\left(\dfrac{2}{15}\right)$ 9. _____

10. (2.5)(–0.4)(–3.2)(5)   10. _____

Divide.

11. (–87) ÷ (–3)   11. _____

12. 87.5 ÷ (–0.5)   12. _____

13. (–100) ÷ 0   13. _____

14. $-\dfrac{11}{16} \div \dfrac{33}{40}$ 14. _____

15. $\dfrac{-49}{-7}$ 15. _____

Name _____   Date _____

Practice Set 1.4
Exponents

Write each product in exponent form. Do not evaluate.

1. $4 \times 4 \times 4 \times 4 \times 4$ 1. _____

2. $10 \times 10 \times 10 \times 10$ 2. _____

3. $(x)(x)(x)(x)(x)(x)$ 3. _____

4. $(-ab)(-ab)$ 4. _____

Evaluate.

5. 10^7 5. _____

6. 2^9 6. _____

7. $\left(\dfrac{3}{7}\right)^2$ 7. _____

8. $(3.2)^2$ 8. _____

Name _____ Date _____

9. $(-5)^3$ 9. _____

10. $\left(-\dfrac{1}{2}\right)^3$ 10. _____

11. $(-7)^3(-2)^2$ 11. _____

12. $2^3 + 4^3$ 12. _____

13. $7^2 + 2^2$ 13. _____

14. $-6^2 - (-2)^2$ 14. _____

15. $2^6 - (-3)^4$ 15. _____

16. $7^2 - (-2)^3$ 16. _____

Name _____   Date _____

Practice Set 1.5
The Order of Operations

Evaluate. Simplify all fractions.

1. $1 + 2 \times 3$ 1. _____

2. $12 - 6 \div 3$ 2. _____

3. $1 + 2 - 3 \times 4 \div 6$ 3. _____

4. $9 \times 3 - 14 \div 2$ 4. _____

5. $4 \times 8 + 9 - 3$ 5. _____

6. $5 \times 2^3 - 4 \times (14 \div 7)$ 6. _____

7. $50 \div 5 \times 2 + (11 - 8)^2$ 7. _____

8. $9^2 + 4^2 \div 2^2 - 4^4 \div 2^2$ 8. _____

19

Name _____   Date _____

9. $12 - 8(3)^2 \div 6$ 9. _____

10. $(-3)^3 \cdot 2 \div 3^2 + 2^2$ 10. _____

11. $\left(\dfrac{1}{2}\right)^2 \div \left(\dfrac{1}{2}\right)^3$ 11. _____

12. $(-0.2)^2 \div \left(\dfrac{1}{10}\right)^2$ 12. _____

13. $\left(\dfrac{2}{3}\right)^2 (-18) + \dfrac{3}{7} \div \dfrac{5}{21}$ 13. _____

14. $\left(\dfrac{1}{3}\right)^2 + \left(\dfrac{1}{3}\right)^2 - \left(-\dfrac{2}{3}\right)^2$ 14. _____

15. $(0.5 + 2.3)^2 \div \left(\dfrac{1}{2}\right)^2$ 15. _____

Name _____   Date _____

Practice Set 1.6
Using the Distributive Property to Simplify Algebraic Expressions

Multiply. Use the distributive property.

1. $3(x + 2)$ 1. _____

2. $4(2x - 5)$ 2. _____

3. $-2(6x + 3)$ 3. _____

4. $-(-8x - 2y)$ 4. _____

5. $2(-6x - 8y - 4z)$ 5. _____

6. $\dfrac{2}{3}(-9x + 6y + 27)$ 6. _____

7. $\dfrac{1}{2}\left(\dfrac{1}{3}x + \dfrac{2}{5}y\right)$ 7. _____

8. $3x(2x - y - z)$ 8. _____

9. $\dfrac{x}{3}(x + 2y + 2z)$ 9. _____

Name _____   Date _____

10. $(4x - 5)(2)$ 10. _____

11. $(6x - 4y + 2z)(-3)$ 11. _____

12. $2.1(3.4x^2 + 0.2)$ 12. _____

13. A rectangular birthday that is 18 inches long is cut into two   13. _____
    pieces. One piece is 5 inches wide, and the other piece is $3x$
    inches wide. Write an expression to represent the area of the
    entire birthday cake in square inches and then simplify this
    expression.

14. Mr. Jorgensen's backyard is a rectangle that is 30 feet wide. He   14. _____
    recently built a fence that extends across his backyard. The
    distance from his house to the fence is 15 feet. The distance from
    the fence to the back edge of his property is $7n$. Write an
    expression that represents the area of Mr. Jorgensen's property in
    square feet, then simplify this expression.

15. Tami is painting a large wall that is 25 feet wide. The wall is cut   15. _____
    in half by a horizontal chair rail that extends from one side of the
    wall to the other. The vertical distance from the floor to the chair
    rail is $3\frac{1}{2}$ feet. The vertical distance from the chair rail to the
    ceiling is $(8a - 5)$ feet. Write and simplify an expression that
    represents the area of the wall in square feet.

Name _____   Date _____

Practice Set 1.7
Combining Like Terms

Combine like terms.

1. $3x^2 + 8x^2$

1. _____

2. $-13b^5 - 5b^5$

2. _____

3. $5a^2 - 16a^2 + 4a^4$

3. _____

4. $-5y + 3z - 2y - 8z$

4. _____

5. $2.1x - 3.2y + 1.4x + 4.1y$

5. _____

6. $6.07y^2 - 3.9y + 2.89y^2$

6. _____

7. $3x - 7y - 2x + z - 4z + 3y$

7. _____

8. $9a + 10 - 2a^2 - 11a + 2 + 6a^2$

8. _____

Name _____   Date _____

9. $\dfrac{3}{5}x - \dfrac{2}{3}y - \dfrac{4}{5}x + \dfrac{1}{3}y$ 9. _____

10. $\dfrac{3}{7}x^2 - \dfrac{1}{2}y + \dfrac{3}{14}x^2 - \dfrac{1}{8}y$ 10. _____

Simplify. Use the distributive property to remove parentheses; then combine like terms.

11. $2(3x - 4) + 5(2x + 4)$ 11. _____

12. $-2(7a - 3b) + 4(-2a + 4b)$ 12. _____

13. $5(6 - b) - 2(-2 - 9b)$ 13. _____

14. $12(5 - y) - 8(12 - 11y)$ 14. _____

15. $2a(b - 4c) - 3c(-7a + b - 10d)$ 15. _____

Name _____   Date _____

Practice Set 1.8
Using Substitution to Evaluate Algebraic Expressions and Formulas

Evaluate.

1. $-3x + 1$ for $x = 5$ 1. _____

2. $-7a - 8$ for $a = 2$ 2. _____

3. $\frac{2}{7}y - 7$ for $y = 14$ 3. _____

4. $4x + 9$ for $x = \frac{1}{3}$ 4. _____

5. $10a - 7$ for $a = \frac{2}{3}$ 5. _____

6. $x^2 - 5x$ for $x = 3$ 6. _____

7. $6 - c^2$ for $c = -5$ 7. _____

8. $2x^2 + 3x - 8$ for $x = 3$ 8. _____

Name _____   Date _____

9. $\frac{1}{2}x^2 - 2x + 8$ for $x = 6$ 9. _____

10. $a^2 + 3ab - 2b^2$ for $a = -2$ and $b = -3$ 10. _____

11. $2x^3 + 3xy^2 - 4y^3$ for $x = 3$ and $y = -2$ 11. _____

12. $\frac{2a^2 - 3b}{b^2}$ for $a = -1$ and $b = 2$ 12. _____

13. A carpenter building a sign starts with a rectangular piece of wood that is 30 inches long and 20 inches wide. Before she begins painting, she cuts six inches off the length and reduces the width by 4 inches. What is the area of the sign in square inches?   13. _____

14. A triangular table top has a base of 60 centimeters and a height of 31 centimeters. What is the area of the table top in square centimeters?   14. _____

15. Roger wants a trapezoidal section of his roof re-shingled. The height of this section is 12 feet, the top base is 8 feet in length, and the bottom base is 14 feet. If the roofer charges $12 per square foot, how much will the job cost?   15. _____

16. A circular patch of grass has a radius of 6 meters. What is the area of the patch of grass, rounded to the nearest square meter?   16. _____

Name _____   Date _____

Practice Set 1.9
Grouping Symbols

Simplify. Remove grouping symbols and combine like terms.

1. $4 + 2(n - 7)$　　　　　　　　　　　　　　　　1. _____

2. $8x - 3(x + 2)$　　　　　　　　　　　　　　　2. _____

3. $-2a - 3(a + 3b)$　　　　　　　　　　　　　　3. _____

4. $3(2a + 4b) - (4a - 3b)$　　　　　　　　　　　4. _____

5. $11y + 5(3x - y) - 6(y - x)$　　　　　　　　　5. _____

6. $2(2a + b) - 5(b - 4) + 9(a - 2)$　　　　　　　6. _____

7. $3x [2x^2 - 3(x + 4)]$　　　　　　　　　　　　7. _____

8. $-2[3(x - y) - 2(x + y)]$　　　　　　　　　　　8. _____

Name _____   Date _____

9. $5[-3(2x + 4y) + 6(-2x + 3y)]$ 9. _____

10. $2[2x - 3(4x - 2y) + 3y^2]$ 10. _____

11. $3b^2 - 2[4b + 3b(-1 - b)]$ 11. _____

12. $2b - \{3a + 4[2a - (-2b + 3a)]\}$ 12. _____

13. $-x - \{5y - 2[x - (2y - 5)]\}$ 13. _____

14. $2\{2x^2 + 3[4y - (2 - y)]\}$ 14. _____

15. $-2\{5x^2 + 2[3x - (5 - x)]\}$ 15. _____

Name _____   Date _____

Practice Set 2.1
The Addition Principle of Equality

Solve for *x*. Check your answers.

1. $x + 6 = 12$ 1. _____

2. $x + 7 = 22$ 2. _____

3. $14 = 8 + x$ 3. _____

4. $0 = x - 8$ 4. _____

5. $x - 8 = -5$ 5. _____

6. $12 + 4 = x - 3$ 6. _____

7. $14 - 6 + x = 27 - 1$ 7. _____

8. $-27 + x - 11 = 40 - 22$ 8. _____

Name _____   Date _____

9. $\dfrac{2}{5} + x = \dfrac{3}{10} + \dfrac{4}{5}$ 9. _____

10. $1.7 + 4.2 + x = 19.23 - 9.8$ 10. _____

Determine whether the given solution is correct. If it is not, find the solution.

11. Is $x = -3$ the solution to $-8 + x = -5$?   11. _____

12. Is $x = 6$ the solution to $x + 7 = 14 - 19 + 6$?   12. _____

13. Is $x = 24$ the solution to $x - 10 = 20 - 11 + 5$?   13. _____

14. Is $x = 28$ the solution to $-17 + x - 4 = 12 - 8 + 3$?   14. _____

15. Is $x = 29$ the solution to $-2 - x + 6 = 18 + 22 - 7$?   15. _____

Name _____   Date _____

Practice Set 2.2
The Multiplication Principle of Equality

Solve for *x*. Be sure to reduce your answer. Check your solution.

1. $11x = 44$

    1. _____

2. $63 = 7x$

    2. _____

3. $-13 = -x$

    3. _____

4. $\dfrac{1}{8}x = 2$

    4. _____

5. $\dfrac{1}{2}x = 11$

    5. _____

6. $-5 = \dfrac{x}{5}$

    6. _____

7. $-25 = 15x$

    7. _____

8. $1.5x = 45$

    8. _____

Name _____   Date _____

9. $-2x - 4x = -42$ 9. _____

10. $14 - 23 = -3x$ 10. _____

Determine whether the given solution is correct. If it is not, find the correct solution.

11. Is $x = 6$ the solution for $8x = 56$? If not, find the correct solution.   11. _____

12. Is $x = 7$ the solution for $\dfrac{3}{7}x = -3$? If not, find the correct solution.   12. _____

13. Is $x = 49$ the solution for $\dfrac{4}{7}x = 28$? If not, find the correct solution.   13. _____

14. Is $x = 0.12$ the solution for $3x = -0.36$? If not, find the correct solution.   14. _____

15. Is $x = 3$ the solution for $8x - 5x = -12$? If not, find the correct solution.   15. _____

Name _____   Date _____

Practice Set 2.3
Using the Addition and Multiplication Principles Together

Find the value of the variable that satisfies the equation. Check your solution. Answers that are not integers may be left in fractional form or decimal form.

1. $3x + 15 = 30$ 1. _____

2. $3x + 19 = -14$ 2. _____

3. $9x - 14 = 22$ 3. _____

4. $5x - 28 = -68$ 4. _____

5. $8x = 2x + 36$ 5. _____

Solve for the variable. You may move the variable terms to the right or to the left.

6. $2x - 3 = x + 9$ 6. _____

7. $7x + 5 = 4x + 23$ 7. _____

Name _____   Date _____

8. $10x - 8 = 6x + 24$　　　　　　　　　　8. _____

9. $11x + 14 = 3x - 14$　　　　　　　　　9. _____

10. $18 - 3 = 16x - 4x + 12$　　　　　　　10. _____

11. $5(x + 3) - 3(x - 2) = 35$　　　　　　11. _____

12. $3(2z - 3) - 3 = 3z - 2(2z - 1)$　　　12. _____

13. $9(2y - 2) + 13 = 3(y + 8) + 16$　　　13. _____

14. $10(n - 1) + 18 = 7(n + 3 - 1)$　　　 14. _____

15. $8(2p - 4) - 9(p + 2) = 2 - 5p$　　　 15. _____

Name _____   Date _____

Practice Set 2.4
Solving Equations with Fractions

Solve for the variable and check your answer. Noninteger answers may be left in fractional form or decimal form.

1. $\dfrac{1}{5}x + \dfrac{2}{5} = \dfrac{4}{5}$ 1. _____

2. $\dfrac{1}{3}p = p - 1$ 2. _____

3. $\dfrac{1}{2}n - \dfrac{3}{4} = \dfrac{3}{4}$ 3. _____

4. $\dfrac{1}{3}x - \dfrac{1}{2} = \dfrac{1}{6}$ 4. _____

5. $\dfrac{1}{2} - \dfrac{1}{8}x = \dfrac{11}{16}$ 5. _____

6. $\dfrac{2}{3}x = \dfrac{1}{21}x + \dfrac{3}{7}$ 6. _____

7. $\dfrac{a-3}{5} = 1 - \dfrac{a}{3}$ 7. _____

8. $\dfrac{x+2}{3} = \dfrac{x}{18} + \dfrac{1}{9}$ 8. _____

Name _____   Date _____

9. $\dfrac{3}{8}x + \dfrac{2}{3} = \dfrac{3x+4}{24}$

9. _____

10. $\dfrac{3x-5}{6} = \dfrac{14-4x}{3}$

10. _____

11. $\dfrac{a+1}{35} = -\dfrac{a}{7} - 1$

11. _____

Remove parentheses first. Then combine like terms. Solve for the variable. Noninteger answers may be left in fractional form or decimal form.

12. $\dfrac{3}{4}(2x+6) - 3 = 3(x+3)$

12. _____

13. $3(-x-1) = -\dfrac{4}{5}(6x+6) + 3$

13. _____

14. $0.4(x-6) = 0.7(2x+3) + 0.5$

14. _____

15. $-4(0.5x + 0.3) - 1.8 = 1.6$

15. _____

Name _____   Date _____

Practice Set 2.5
Translating English Phrases into Algebraic Expressions

Write an algebraic expression for each quantity. Let x represent the unknown value.

1. seven greater than a number   1. _____

2. a value increased by 12   2. _____

3. eighteen fewer than a quantity   3. _____

4. one-seventh of a quantity   4. _____

5. one-fourth of a number, decreased by one   5. _____

6. four more than half of a number   6. _____

7. sixteen less than three times a number   7. _____

8. ten more than twice a number   8. _____

Name _____   Date _____

9. one-half of the sum of a number and two   9. _____

10. the sum of a number and two divided by five   10. _____

Write an algebraic expression for each of the quantities being compared.

11. The value of Alicia's car is $2300 more than the value of Allison's car.   11. _____

12. The length of the rectangle is 14 yards more than twice the width.   12. _____

13. The cost of dinner for Raymond was $4 more than $\frac{1}{3}$ the cost of John's.   13. _____

14. A recent survey on campus revealed that the number of business majors was forty more than the number of nursing students. The number of criminal justice majors was 14 more than two-thirds the number of nursing students.   14. _____

15. Patricia owns 17 more comic books than Scott, Walter owns three times as many as Scott, and Adrienne owns five fewer than four times as many as Scott.   15. _____

Name _____   Date _____

Practice Set 2.6
Using Equations to Solve Word Problems

Solve. Check your solution.

1. What number minus 312 gives 234?   1. _____

2. A number divided by 13 is 14. What is the number?   2. _____

3. A number is tripled and then increased by 27. The result is 72. What is the original number?   3. _____

4. Twice a number is increased by three-fifths the same number. The result is fifty-two. Find the number.   4. _____

5. The sum of a number, one-fifth of that number, and twice that number is forty-eight. What is the original number?   5. _____

6. Sixteen less than ten times a number is the same as twice the number. Find the number.   6. _____

7. Half of a number plus 15 is the same as three times the number minus five. What is the number?   7. _____

8. A number minus one-fourth of that number plus three times that number added together gives you 11 more than the original number. Find the number.   8. _____

Name _____   Date _____

9. If you start with a number, add 3, multiply the result by 9, and add 1, you end up with 13 times the number you started with. What is the original number?

9. _____

Solve. Check to see if your answer is reasonable.

10. The local health food store has six times as many energy drinks as energy bars. There are 108 energy drinks in stock. How many energy bars are in stock?

10. _____

11. A computer shop has three times as many new computers as used computers. If there are a total of 84 of both kinds of computers, how many used computers does the computer shop have?

11. _____

12. Every week, Jeff gets paid $12 per hour up to 40 hours and $18 per hour for every hour over 40. Last week, he earned $642. How many hours did he work?

12. _____

13. Anthony and Sarah both work at a telephone market research company. Yesterday, Anthony filled out $\frac{3}{4}$ as many surveys as Sarah, and together they filled out 35 surveys. How many surveys did Anthony fill out?

13. _____

14. Amy's cell phone company charges $10.50 per month for 200 minutes of use and $0.15 for each additional minute. Last month Amy's cell phone bill was $55.50. How many additional minutes was she charged for?

14. _____

15. Beth is in training for a marathon. Over the last three months, she has run 367.2 miles in exactly 72 hours. Assuming a constant rate, what is Beth's speed as she runs?

15. _____

Name _____   Date _____

Practice Set 2.7
Solving Word Problems Using Geometric Formulas

Solve the following problems. Use 3.14 for π.

1. What is the perimeter of a countertop whose two sides are 16 feet and 4 feet?

    1. _____

2. What is the area of a parallelogram with a base of 12 cm and an altitude of 5 cm?

    2. _____

3. Find the area of a circular mirror whose diameter is 23 inches. Round to the nearest square inch.

    3. _____

4. A triangle with an area of 120 m^2 has a base of 16 meters. What is the altitude of the triangle?

    4. _____

5. What are the surface area and volume of a sphere with a radius of 22 inches? Round to the nearest square inch or cubic inch.

    5. _____

6. Find the altitude of a triangular window whose base is 22 inches and area is 396 in^2.

    6. _____

7. The area of a trapezoid is 576 m^2. The altitude is 24 meters and one of its bases is 18 meters. Find the length of the other base.

    7. _____

8. The radius of the Earth is approximately 4000 miles. Use this approximation to find the surface area of the Earth, assuming that it is a perfect sphere.

    8. _____

Name _____   Date _____

9. Find the area of a circular table whose circumference is 18.84 feet.   9. _____

10. The perimeter of a trapezoid is 25 inches. The two sides and the shorter base are all the same length, while the longer base is twice this length. What is the length of the longer base?   10. _____

11. The smallest angle of a triangle measures one-third of the largest angle and one-half of the remaining angle. What is the measurement of the smallest angle?   11. _____

12. The Lorimers want to put a fence around a rectangular area of land for their six horses. They want the width of this area to be 360 feet. If each horse needs an acre (43,560 square feet) of land to graze upon, how long will the entire fence be?   12. _____

13. If a bag of grass seed can cover 225 ft^2, how many bags of grass seed are needed to cover a rectangular area whose width is 45 feet and length is 90 feet?   13. _____

14. A rectangular box is manufactured to hold gourmet candy. The box will measure 5 inches in length, 2 inches in width, and 2 inches in height. If lightweight cardboard costs $0.03 per square inch, how much will it cost to produce 325 candy boxes?   14. _____

15. A cylindrical storage container is 10 meters tall and has a capacity of 1,538.6 cubic meters. If you walk at a rate of 2 meters per second, how much time (to the nearest whole second) will you need to walk all the way around the container?   15. _____

Name _____   Date _____

Practice Set 2.8
Solving Inequalities in One Variable

Replace the ? by < or >.

1. $13 \; ? \; -2$ 1. _____

2. $-10 \; ? \; -7$ 2. _____

3. $-\dfrac{2}{3} \; ? \; -\dfrac{3}{4}$ 3. _____

Graph each inequality on the number line.

4. $x < -4$ 4. ⊢┼┼┼┼┼┼┼┼┼┼┼→ x

5. $3.8 \leq x$ 5. ⊢┼┼┼┼┼┼┼┼┼┼┼→ x

Solve and graph the result.

6. $x + 5 \leq 13$ 6. _____   ⊢┼┼┼┼┼┼┼┼┼┼┼→ x

7. $x - 11 < -25$ 7. _____   ⊢┼┼┼┼┼┼┼┼┼┼┼→ x

8. $6x \leq 36$ 8. _____   ⊢┼┼┼┼┼┼┼┼┼┼┼→ x

Name _____   Date _____

9. $-3x < -39$ 9. _____ ———————————→ x

10. $\frac{1}{5}x \leq 4$ 10. _____ ———————————→ x

11. $-\frac{2}{3}x > -8$ 11. _____ ———————————→ x

12. $3x + 5 < 14$ 12. _____ ———————————→ x

13. $5y - 4 \geq -6$ 13. _____ ———————————→ x

Solve. Collect the variable terms on the left side of the inequality.

14. $-3 + 2x < 6x + 9$ 14. _____

15. $4(2x + 6) > 3(2x + 2)$ 15. _____

16. $\frac{3x}{4} + 3 \leq \frac{2x}{3} + 5$ 16. _____

44

Name _____   Date _____

Practice Set 3.1
The Rectangular Coordinate System

Plot the following points.

1. A: (5, 1)

2. B: (−3, 4)

3. C: (2, −4)

4. D: (−1, −3)

5. E: (3.5, −2)

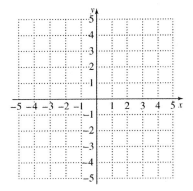

Consider the points plotted in the graph below. Give the coordinates for the points listed.

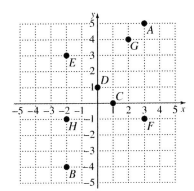

6. A   6. _____

7. B   7. _____

8. E   8. _____

9. F   9. _____

10. H   10. _____

45

Name _____ Date _____

Find the missing coordinate to complete the ordered-pair solution to the given linear equation.

11. $y = 3x - 2$
    a. (0, ?)
    b. (2, ?)

11. a._____

    b._____

12. $y = -2x + 3$
    a. (−1, ?)
    b. (3, ?)

12. a._____

    b._____

13. $3x - y = 4$
    a. (0, ?)
    b. (−1, ?)

13. a._____

    b._____

14. $-2x + 3y = -12$
    a. (3, ?)
    b. (0, ?)

14. a._____

    b._____

15. $4x - 2y = 12$
    a. (−1, ?)
    b. (4, ?)

15. a._____

    b._____

Name _____   Date _____

Practice Set 3.2
Graphing Linear Equations

Graph each equation by plotting three points and connecting them.

1. $y = -3x + 1$

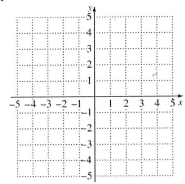

2. $y = 2x - 5$

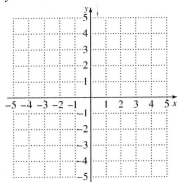

3. $y = \dfrac{1}{2}x - 4$

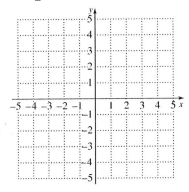

4. $y = -\dfrac{1}{2}x + 2$

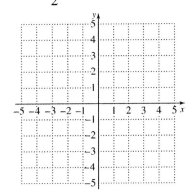

5. $5x + 2y = 10$

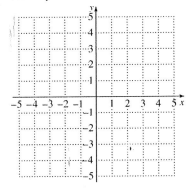

6. $2x - 6 = -3y$

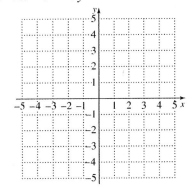

Name _____   Date _____

Graph each equation by plotting the intercepts and one other point.

7. $y = 5 - x$

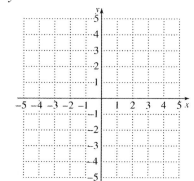

8. $y = 4 - 2x$

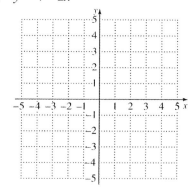

9. $x - 4 = 2y$

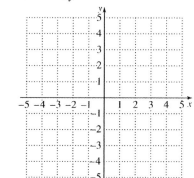

10. $x + 3 = 3y$

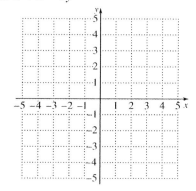

11. $2y = 4 - x$

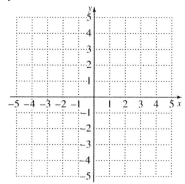

12. $2x - 5 = y$

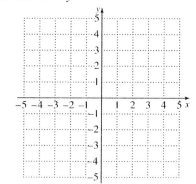

Name _____   Date _____

Practice Set 3.3
The Slope of a Line

Find the slope of the straight line that passes through the given pair of points.

1. (5, 2) and (8, 5)   1. _____

2. (10, 7) and (−2, 1)   2. _____

3. (−5, −2) and (1, −4)   3. _____

4. (3, −5) and (2, −9)   4. _____

Find the slope and the *y*-intercept.

5. $y = \dfrac{3}{4}x - 4$ 5. _____

6. $y = 5x$ 6. _____

7. $x = -3y + 6$ 7. _____

Write the equation of the line in slope-intercept form.

8. $m = \dfrac{2}{3}$, *y*-intercept (0, 4)   8. _____

9. $m = -4$, *y*-intercept $\left(0, -\dfrac{4}{5}\right)$ 9. _____

Name _____   Date _____

10. $m = 0$, y-intercept $(0, -2)$ 10. _____

11. $m = \dfrac{6}{5}$, y-intercept $(0, 4)$ 11. _____

Graph the line $y = mx + b$ for the given values.

12. $m = -\dfrac{3}{4}, b = 4$

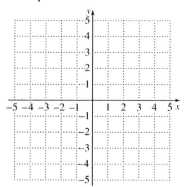

13. $m = \dfrac{2}{3}, b = -3$

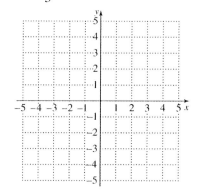

14. A line has a slope of $\dfrac{1}{4}$.   14. a._____
   a. What is the slope of the line parallel to it?
   b. What is the slope of the line perpendicular to it?   b._____

15. A line has a slope of 3.   15. a._____
   a. What is the slope of the line parallel to it?
   b. What is the slope of the line perpendicular to it?   b._____

16. The equation of a line is $y = 6x - 3$.   16. a._____
   a. What is the slope of the line parallel to it?
   b. What is the slope of the line perpendicular to it?   b._____

Name _____   Date _____

Practice Set 3.4
Writing the Equation of a Line

Find an equation of the line that has the given slope and passes through the given point.

1. $m = 3, (-2, -1)$ 1. _____

2. $m = -4, (2, 1)$ 2. _____

3. $m = -1, (0, 4)$ 3. _____

4. $m = \dfrac{2}{3}, (6, 7)$ 4. _____

5. $m = -\dfrac{1}{2}, (3, 1)$ 5. _____

6. $m =$ undefined, $(3, 0)$ 6. _____

Write an equation of the line passing through the given points.

7. $(-1, 2)$ and $(5, -1)$ 7. _____

8. $(-3, 2)$ and $(1, 10)$ 8. _____

Name _____   Date _____

9. (3, 4) and (−1, −16)   9. _____

10. $\left(1, \frac{1}{6}\right)$ and $\left(2, \frac{4}{3}\right)$ 10. _____

11. (−3, 0) and $\left(\frac{1}{3}, \frac{4}{3}\right)$ 11. _____

12. (−1, −24) and (−8, −10)   12. _____

13. Find the equation of a line with a slope of −3 that passes through the point (−2, 4).   13. _____

14. Find the equation of a line that passes through (−6, −9) and has zero slope.   14. _____

15. Find the equation of a line that passes through (3, −7) and is parallel to $y = -5x + 2$.   15. _____

Name _____ Date _____

Practice Set 3.5
Graphing Linear Inequalities

Graph the region described by the inequality.

1. $y < 2x + 1$

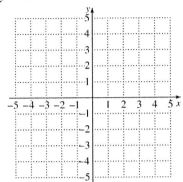

2. $y > -3x - 5$

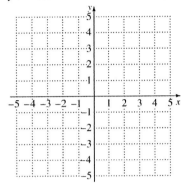

3. $y \leq -3x$

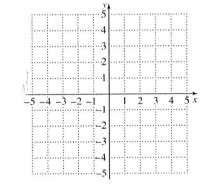

4. $3x - 5y - 10 \geq 0$

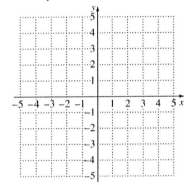

5. $3x \leq 5y$

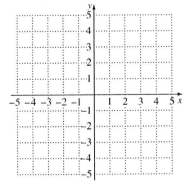

6. $-2x > 1 + y$

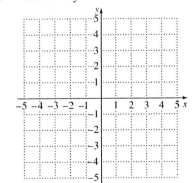

Name _____ Date _____

7. $y \geq -\dfrac{1}{2}x$

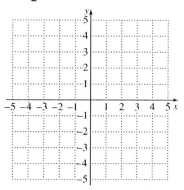

8. $y < -\dfrac{1}{2}x + 4$

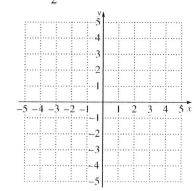

9. $x > -3$

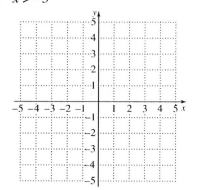

10. $y \leq 4$

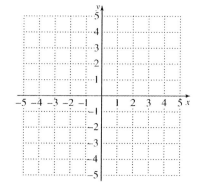

11. $2x - 3y + 9 \geq 0$

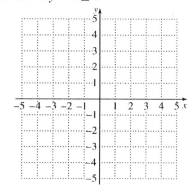

12. $3x + 6y - 9 < 0$

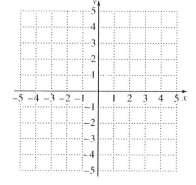

Name _____   Date _____

Practice Set 3.6
Functions

Find the domain and range of the relation. Determine whether the relation is a function.

1. $\{(2,3),(\frac{1}{2},4),(-4,-3),(5,-7)\}$

    1. a._____

       b._____

2. $\{(\frac{4}{5},4),(0,\frac{4}{5}),(\frac{4}{5},\frac{1}{2}),(4,0)\}$

    2. a._____

       b._____

3. $\{(2.5,3),(3.5,0),(5.5,-2),(8.5,-6)\}$

    3. a._____

       b._____

4. $\{(3,77),(5,79),(3,81),(8,83)\}$

    4. a._____

       b._____

Graph the equation.

5. $y = x^2 + 2$

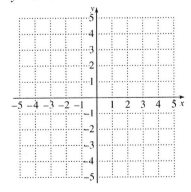

6. $x = y^2 - 3$

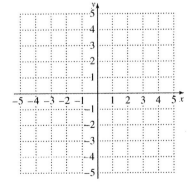

55

Name _____   Date _____

7. $y = -3x^2$

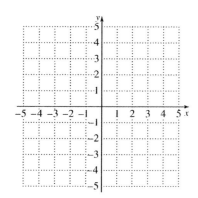

8. $y = \dfrac{4}{x}$

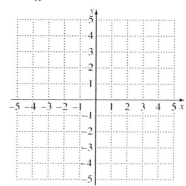

Determine whether each relation is a function.

9. _____

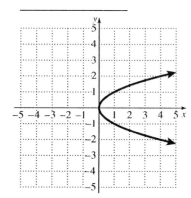

10. _____

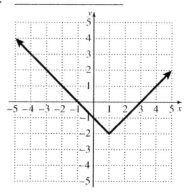

Given the following functions, find the indicated values.

11. $f(x) = 4 - 3x$
    a. $f(2)$
    b. $f(-3)$
    c. $f(0)$

11. a._____

    b._____

    c._____

12. $f(x) = 2x - 5$
    a. $f(1)$
    b. $f(-2)$
    c. $f(4)$

12. a._____

    b._____

    c._____

13. $f(x) = 2x^2 - 3x + 1$
    a. $f(0)$
    b. $f(-3)$
    c. $f(3)$

13. a._____

    b._____

    c._____

Name _____   Date _____

Practice Set 4.1
Systems of Linear Equations in Two Variables

Solve the system of equations by graphing. Check your solution.

1. $-x + y = -2$
    $x + 3y = 18$

1. 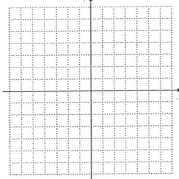 _____

2. $4x - y = 4$
    $x + 2y = 10$

2. 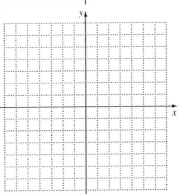 _____

3. $5x - 2y = 6$
    $y = \dfrac{5}{2}x + 1$

3. 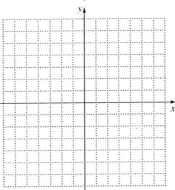 _____

4. $3y + 2x = 9$
    $6y = -4x + 18$

4. 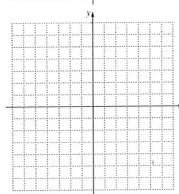 _____

57

Name _____ Date _____

If possible, solve each system of equations. Use any method. If there is not a unique solution to a system, state a reason.

5. $3x + y = -1$
    $4x + 2y = 0$

    5. _____

6. $y = -\dfrac{1}{2}x + 1$
    $2x + 5y = 1$

    6. _____

7. $y = 3x + 2$
    $-15x + 5y = 0$

    7. _____

8. $x + y = -10$
    $x - y = 0$

    8. _____

9. $5x = y$
    $3x - y = -4$

    9. _____

10. $6x + 3y = 12$
    $6x - 3y = -12$

    10. _____

11. $10y - 2x = 8$
    $y = \dfrac{1}{5}x + \dfrac{4}{5}$

    11. _____

12. $0.4x + 0.5y = 0.3$
    $-10y = 8x + 8$

    12. _____

Practice Set 4.2
Systems of Linear Equations in Three Variables

Solve each system.

1. $x + y + z = 7$
   $2x + 3y - z = -3$
   $x - 6y + z = 0$

2. $x + 4y - z = -15$
   $2x - y - 2z = -12$
   $3x - y + z = 1$

3. $x + y + z = 5$
   $x - y - z = -3$
   $2x - 2y + 3z = -6$

4. $x + y = -5 - z$
   $3y + z = -2x - 2$
   $3x + 5y + 2z = -4$

5. $x + y + z = 10$
   $x - y + z = -4$
   $x + y - z = 8$

6. $4y = 3x - z + 13$
   $y + 5z = -2x - 3$
   $3x = y - 3z - 9$

7. $0.2x + 0.4y + 0.6z = 0$
   $0.1x + y - 0.5z = -1.6$
   $0.2x - 0.5y + 0.1z = 0.4$

8. $x + y = 22$
   $x + z = 0$
   $x + y + z = 12$

Name _____ Date _____

9. $x + 3y - z = 6$
    $x + 3z = 0$
    $x - 6y = -1$

9. _____

10. $x + y + z = 0$
    $x + 2y = -3$
    $x + 6 = z - 6$

10. _____

11. $x + 3y = -2$
    $y + z = 2$
    $3x = 2y + 2z + 8$

11. _____

12. $x + 2z = 4$
    $2x + 2y + z = -8$
    $y + 3z = 1$

12. _____

13. $x - y = 0$
    $y - z = 5$
    $x + y + z = 13$

13. _____

14. $0.2x + 0.3y = -1.5$
    $0.1y + 0.3z = 0.3$
    $0.5x - 0.4y + 0.3z = 0.3$

14. _____

Try to solve the system of equations. Explain your result in each case.

15. $2y - z = 8$
    $3x + y = 2$
    $-3x + y - z = 6$

15. _____

16. $x + y - z = 4$
    $2x - 5y + z = 1$
    $3x + 3y - 3z = 0$

16. _____

Name _____   Date _____

Practice Set 4.3
Applications of Systems of Linear Equations

Use a system of two linear equations to solve each exercise.

1. The sum of two numbers is 102. If three times the smaller number is subtracted from twice the larger number, the result is 49. Find the two numbers.

   1. _____

2. The difference between two numbers is 13. If three times the larger number is added to four times the smaller number, the result is 74. Find the two numbers.

   2. _____

3. A carpenter worked for 3 hours on a project, and her helper worked 4 hours. The carpenter charged $144 for the project. Later, the carpenter worked for 5 hours and the helper worked for 8 hours on another project, and the bill was $256. How much does the carpenter charge per hour for her work? How much does she charge per hour for her helper?

   3. _____

4. For a performance of *King Lear*, student tickets cost $18 and adult tickets cost $28. In all, 320 people attended the performance, and ticket sales receipts totaled $8060. How many tickets of each type were sold?

   4. _____

5. The Revel family farm has 500 acres of land. It costs $60 to plant an acre of soybeans and $36 to plant an acre of corn. If the Revels want to spend a total of $22,440 on planting, how many acres of each crop should they plant?

   5. _____

6. A manufacturing plant makes economy microwaves and luxury microwaves. It takes 2.5 minutes to make an economy microwave and 3.5 minutes to make a luxury microwave. The economy microwave requires $8 worth of materials and the luxury microwave requires $15 worth of materials. Yesterday, the plant made microwaves for a total of 6 hours and used $1418 worth of materials. How many microwaves of each type were made yesterday?

   6. _____

7. On Monday morning, a contractor bought 5 bagels and 8 coffees for his workers. He paid $9.70. On Tuesday, the contractor bought 7 bagels and 7 coffees, paying $10.85. How much does one bagel cost? How much does one coffee cost?

   7. _____

Name _____   Date _____

8. A publishing company is getting estimates for the cost of printing a novel. One printer has a fixed charge to set up the printing and a per-copy cost for each book printed. The printer would charge $19,600 for 3000 books and $28,600 for 4500 books. What is the fixed charge to set up the printing? What is the per-copy cost for printing a book?

8. _____

9. Against the wind, a plane traveled 540 miles in 3 hours. With the wind, the return trip took 2.25 hours. What was the speed of the wind? What was the speed of the plane in still air?

9. _____

10. Eric and Dawn canoe 8 miles upstream in 4 hours to reach a campsite. The next day, they make the return trip in 3.2 hours. How fast can Eric and Dawn canoe in still water? How fast is the current?

10. _____

11. A basketball player scored 25 points in a game without making any three-point shots. He scored a total of 16 times, making several free throws worth 1 point each and several regular shots worth 2 points each. How many free throws did he make? How many 2-point shots did he make?

11. _____

Use a system of three linear equations to solve exercises 12–14.

12. Devon bought 15 items at the office supply store. She spent a total of $26.75. The binders cost $2.40, the pens cost $1.85, and the erasers cost $0.60. Devon bought 4 more pens than erasers. How many of each item did she buy?

12. _____

13. A total of 250 people attended a movie. The tickets cost $11 for adults, $8 for students, and $7 for senior citizens. The ticket sales totaled $2318. The manager found that if they had raised the prices to $14 for adults, $10 for students, and $8 for senior citizens, they would have made $2892. How many tickets of each type were sold?

13. _____

14. An ice cream parlor sells small, medium, and large ice cream cones. The small cone contains 5 ounces of ice cream and costs $2, the medium cone holds 7 ounces of ice cream and costs $3, and the large cone holds 10 ounces of ice cream and costs $4. Last night, the parlor sold 110 ice cream cones containing a total of 47 pounds 6 ounces of ice cream, and collected $311. How many of each size of ice cream cone did the parlor sell?

14. _____

Name _____   Date _____

Practice Set 4.4
Systems of Linear Inequalities

Graph the solution of each of the following systems.

1. $y > 3x - 2$
   $y \geq -\dfrac{1}{2}x + 3$

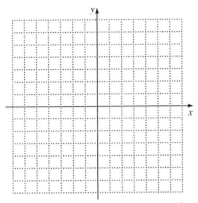

2. $y < x + 1$
   $y < -\dfrac{1}{3}x - 2$

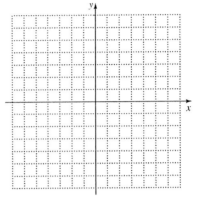

3. $4x + 2y \geq -4$
   $y > x - 3$

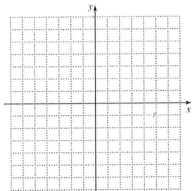

4. $y > -2$
   $y \leq -3x + 2$

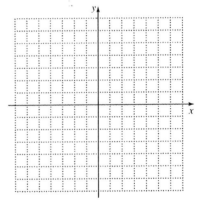

5. $y < 2$
   $x < 5$

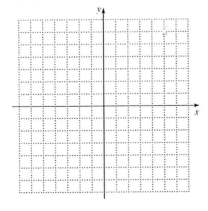

6. $x < -3$
   $y \geq -5$

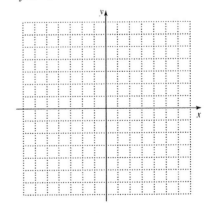

Name _____   Date _____

7. $y + x \leq 4$
    $y + x > 1$

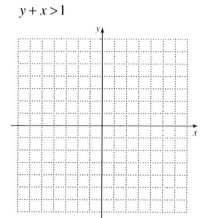

8. $3x - 2y < 6$
    $3x - 2y > -9$

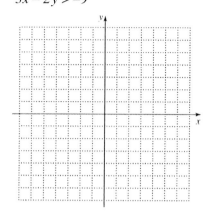

Graph the solution of the following systems of inequalities. Find the vertices of the solution.

9. $y < x$
    $x - 3y \geq 6$

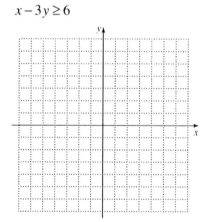

10. $x + y \leq 5$
    $y \leq 2x + 2$

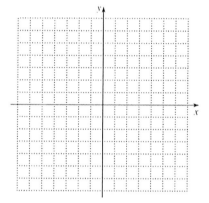

11. $y < x - 2$
    $3y \geq -x + 6$
    $x < 6$

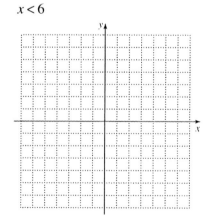

12. $y > -3x - 5$
    $y < 4$
    $-2x + y > 0$

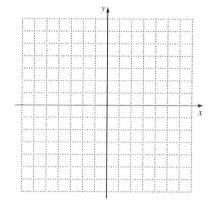

Name _____   Date _____

Practice Set 5.1
The Rules of Exponents

Multiply or divide. Leave your answer in exponent form.

1. $3^2 \cdot 3^5$

1. _____

2. $x^4 \cdot x^7$

2. _____

3. $z^{16} \cdot z^4$

3. _____

4. $(3a^4)(2a^6)$

4. _____

5. $(2a^6b^2)(4ab^3)$

5. _____

6. $(12x^4y^2)(2x^2y^3)$

6. _____

7. $(-13a^2b^4c)(4ab^3c^2)(0)$

7. _____

8. $(3a^2b)(-\dfrac{1}{2}a^3c^2)(-20abc^3)$

8. _____

Name _____ Date _____

Divide. Leave your answer in exponent form. Assume that all variables in any denominator are nonzero.

9. $\dfrac{x^{12}}{x^5}$ 9. _____

10. $\dfrac{y^3}{y^8}$ 10. _____

11. $\dfrac{24x^8 y^3}{-48x^2 y^9}$ 11. _____

12. $\dfrac{9a^3}{a^3}$ 12. _____

13. $\left(\dfrac{2x^2 y^0}{14x^5}\right)^2$ 13. _____

14. $\left(\dfrac{b}{b^3}\right)^4$ 14. _____

15. $\left(-xy^3\right)^2$ 15. _____

Name _____   Date _____

Practice Set 5.2
Negative Exponents and Scientific Notation

Simplify. Express your answer with positive exponents. Assume that all variables are nonzero.

1. x^{-5} 1. _____

2. 7^{-3} 2. _____

3. $\dfrac{1}{2^{-4}}$ 3. _____

4. $\dfrac{x^{-5}y^{-2}}{z^{-3}}$ 4. _____

5. $(9x^2y^{-3})(2x^{-3}y^{-2})$ 5. _____

6. $(2a^3b^{-2}c^{-3})(3a^{-3}b^2c^3)$ 6. _____

7. $(2x^{-2}y^4)^{-5}$ 7. _____

8. $\left(\dfrac{3a^3b^{-2}}{c^3}\right)^{-4}$ 8. _____

Name _____ Date _____

Write in scientific notation.

9. 523

9. _____

10. 7,230,000

10. _____

11. 0.047

11. _____

Write in decimal notation.

12. 6.34×10^3

12. _____

13. 1.37×10^{-4}

13. _____

Evaluate by using scientific notation and the laws of exponents. Leave your answer in scientific notation.

14. (23,000,000)(19,000,000,000)

14. _____

15. $\dfrac{0.0046}{0.023}$

15. _____

Name _____   Date _____

Practice Set 5.3
Fundamental Polynomial Operations

State the degree of the polynomial and whether it is a monomial, a binomial, or a trinomial.

1. $8a^2b^3$

   1. _____

2. $23x^6 - 14x^3 + 5$

   2. _____

3. $5x^2y^4 + 4x^5y^3$

   3. _____

Add.

4. $(-2x + 5) + (8x + 2)$

   4. _____

5. $(3x^2 - 4x + 5) + (-7x^2 + 8x - 12)$

   5. _____

6. $\left(\frac{1}{3}x^2 + \frac{1}{2}x - 6\right) + \left(\frac{1}{6}x^2 - \frac{1}{4}x - 2\right)$

   6. _____

7. $(6.4x - 3) + (4.4x - 11)$

   7. _____

8. $(1.2x^2 - 3.6x - 7) + (10.2x^3 - 5.6x^2 + 7.3x - 9)$

   8. _____

Name _____   Date _____

Subtract.

9. $(7x - 3) - (4x - 9)$ 9. _____

10. $(12x^2 - 3x - 7) - (5x^2 + 7x - 9)$ 10. _____

11. $\left(\dfrac{1}{4}a^2 - \dfrac{5}{6}a + 11\right) - \left(\dfrac{2}{3}a^2 - \dfrac{3}{5}a - 6\right)$ 11. _____

12. $\left(2r^4 - 3r^2 + 14\right) - \left(-3r^4 - 2r^2 + 6\right)$ 12. _____

13. $(3.6x^2 - 2.3x + 4) - (2.3x^3 - 6.1x^2 + 1.7x - 6)$ 13. _____

Buses sold in the U.S. have become more fuel efficient over the years. The number of miles per gallon obtained by a certain type of bus can be described by the polynomial

$$0.37x + 5.31,$$

where x is the number of years since 1970. Use this polynomial to estimate the number of miles per gallon obtained by this type of bus in

14. 1980   14. _____

15. 1995   15. _____

Name _____   Date _____

Practice Set 5.4
Multiplying Polynomials

Multiply.

1. $-4y(3y^3 - 5)$

    1. _____

2. $5x(-2x^3 + 3x)$

    2. _____

3. $3x^3(-2x^2 + 4x - 7)$

    3. _____

4. $2y^2(4y^3 + y^2 - 7y)$

    4. _____

5. $(5x^4 + 3x^3 - 2x^2 - x + 9)(2x^3)$

    5. _____

6. $(-4y^3 - 8y^2 - 7y + 9)(-3y^6)$

    6. _____

7. $(x + 5)(x + 4)$

    7. _____

Name _____   Date _____

8. $(x+3)(x-7)$ 8. _____

9. $(x-3)(x-11)$ 9. _____

10. $(2x-3)(x+6)$ 10. _____

11. $(4y+9)(3y-8)$ 11. _____

12. $(3x-7y)(5x+8y)$ 12. _____

13. $(9x-2)(3x+1)$ 13. _____

14. $(2b+6c)(7b+c)$ 14. _____

15. $(2x^2-3y^3)(4x^2+5y^3)$ 15. _____

Name _____   Date _____

Practice Set 5.5
Multiplication: Special Cases

Multiply. Use the special formulas that apply.

1. $(x-3)(x+3)$

    1. _____

2. $(y-2)(y+2)$

    2. _____

3. $(x+10)(x-10)$

    3. _____

4. $(2x+3)(2x-3)$

    4. _____

5. $(4a-7)(4a+7)$

    5. _____

6. $(5a+3b)(5a-3b)$

    6. _____

7. $(0.2x+8)(0.2x-8)$

    7. _____

Name _____ Date _____

8. $(x-4)^2$ 8. _____

9. $(4x+3)^2$ 9. _____

10. $(7a-5)^2$ 10. _____

Multiply.

11. $(x^2-3x+5)(x+5)$ 11. _____

12. $(8x^2-2x+3)(3x+1)$ 12. _____

13. $(x^2-2x+1)(x^2-3x+4)$ 13. _____

14. $(y^2-2y+3)(y^2-5y-6)$ 14. _____

15. $(x+2)(x-1)(x-5)$ 15. _____

Name _____   Date _____

Practice Set 5.6
Dividing Polynomials

Divide.

1. $\dfrac{24y^4 - 12y^3 + 36y}{6y}$ 1. _____

2. $\dfrac{12a^7 - 4a^5 + 8a^3 - 2a^2}{2a^2}$ 2. _____

3. $\dfrac{48x^9 - 16x^7 + 32x^4}{8x^3}$ 3. _____

4. $(21x^7 - 42x^6 - 7x^3) \div 7x^2$ 4. _____

5. $(12y^4 - 18y^3 + 27y^2) \div 3y^2$ 5. _____

Divide. Check your answers by multiplication.

6. $\dfrac{8x^2 + 26x + 21}{2x + 3}$ 6. _____

7. $\dfrac{10x^2 - 7x - 12}{2x - 3}$ 7. _____

Name _____ Date _____

8. $\dfrac{3x^3 - 5x^2 + 7x - 5}{x - 1}$

8. _____

9. $\dfrac{x^3 - 11x^2 + 30x - 8}{x - 4}$

9. _____

10. $\dfrac{x^3 - 7x^2 + 13x - 15}{x - 5}$

10. _____

11. $\dfrac{x^3 - x^2 - 27}{x - 3}$

11. _____

12. $\dfrac{45x^3 - 27x^2 + 13x - 3}{3x - 1}$

12. _____

13. $\dfrac{2x^3 + 7x^2 + 4x - 1}{2x - 1}$

13. _____

14. $\dfrac{y^3 - 2y - 4}{y - 1}$

14. _____

15. $\dfrac{2x^3 - x^2 - 6}{x + 2}$

15. _____

Name _____ Date _____

Practice Set 6.1
Removing a Common Factor

Remove the largest possible common factor. Check your answers by multiplication.

1. $6x - 3y$ 1. _____

2. $11x + 22y$ 2. _____

3. $4x^2 + 4x$ 3. _____

4. $3x^3 + 12x^2 - 21x$ 4. _____

5. $16ab - 12ab^2 - 10ab^3$ 5. _____

6. $15b^5 + 25b^4 - 20b^3$ 6. _____

7. $60x^2y + 18xy - 24x$ 7. _____

Name _____   Date _____

8. $24a^2b + 15ac^2 - 12b^2c$ 8. _____

9. $9x^3y^2 + 18x^2y^2$ 9. _____

10. $26xy^2z^2 + 13x^2yz^3$ 10. _____

11. $8a(x + 3y) - b(x + 3y)$ 11. _____

12. $6(2r + s) - z(2r + s)$ 12. _____

13. $7x(y - 3) + 5(y - 3)$ 13. _____

14. $8a(3x - 2y) + 9b(3x - 2y)$ 14. _____

15. $5a(4x - 3) - (4x - 3)$ 15. _____

Practice Set 6.2
Factoring by Grouping

Factor by grouping. Check your answers.

1. $y(y-2) + 4(y-2)$

    1. _____

2. $x(x+1) - 2(x+1)$

    2. _____

3. $z(z-1) + 3(z-1)$

    3. _____

4. $x^2 + 3x + 4x + 12$

    4. _____

5. $x^2 - 2x + 5x - 10$

    5. _____

6. $2x^2 - 2x + 5x - 5$

    6. _____

7. $xy - 2x + 5y - 10$

    7. _____

Name _____   Date _____

8. $xy - 3y + 7x - 21$ 8. _____

9. $xy - 4xz - y + 4z$ 9. _____

10. $z^2 - 3z + 5z - 15$ 10. _____

11. $6x^2 + 15x - 4x - 10$ 11. _____

12. $8ax - 2a + 4bx - b$ 12. _____

13. $10x^2 + 35x - 12x - 42$ 13. _____

14. $ax - xb - ay + by$ 14. _____

15. $12ax + 15ay + 8xb + 10by$ 15. _____

Name _____ Date _____

Practice Set 6.3
Factoring Trinomials of the Form $x^2 + bx + c$

Factor.

1. $x^2 + 4x + 4$ 1. _____

2. $x^2 + 4x + 3$ 2. _____

3. $x^2 + 6x + 8$ 3. _____

4. $x^2 - 5x + 4$ 4. _____

5. $x^2 - 8x + 12$ 5. _____

6. $a^2 - 7a + 12$ 6. _____

7. $x^2 - 13x + 30$ 7. _____

Name _____   Date _____

8. $x^2 - 8x + 7$ 8. _____

9. $a^2 - 5a - 14$ 9. _____

10. $x^2 - x - 12$ 10. _____

11. $a^2 - 4a - 5$ 11. _____

12. $x^4 + 11x^2 + 28$ 12. _____

13. $y^4 + 11y^2 + 30$ 13. _____

14. $2x^2 + 18x + 28$ 14. _____

15. $3x^2 + 9x - 84$ 15. _____

Name _____ Date _____

Practice Set 6.4
Factoring Trinomials of the Form $ax^2 + bx + c$

Factor by the trial-and-error method. Check your answers by using FOIL.

1. $2x^2 + 9x + 7$
   1. _____

2. $5x^2 + 6x + 1$
   2. _____

3. $4x^2 - 12x + 5$
   3. _____

Factor by the grouping number method. Check your answers by using FOIL.

4. $2x^2 - 5x - 3$
   4. _____

5. $3x^2 - 14x - 5$
   5. _____

6. $7x^2 - 4x - 3$
   6. _____

Factor by any method.

7. $4x^2 - 4x - 3$
   7. _____

Name _____   Date _____

8. $3x^2 + 8x + 5$

8. _____

9. $6y^2 + y - 2$

9. _____

10. $5x^2 + 43x - 18$

10. _____

11. $7x^2 - 13x - 2$

11. _____

12. $5y^2 + 17y + 6$

12. _____

Factor by first factoring out the greatest common factor.

13. $12x^2 - 14x - 40$

13. _____

14. $18x^2 - 6x - 4$

14. _____

15. $15y^2 + 51y - 36$

15. _____

Name _____   Date _____

Practice Set 6.5
Special Cases of Factoring

Factor by using the difference-of-two-squares formula.

1. $16x^2 - 1$ 1. _____

2. $4y^2 - 49$ 2. _____

3. $81x^2 - 25$ 3. _____

4. $25a^2 - 36b^2$ 4. _____

5. $81y^2 - 1$ 5. _____

Factor by using the perfect-square trinomial formula.

6. $4x^2 + 12x + 9$ 6. _____

7. $9y^2 + 12y + 4$ 7. _____

85

Name _____   Date _____

8. $y^2 - 14y + 49$ 8. _____

9. $16y^2 - 24y + 9$ 9. _____

10. $36a^2 + 60ab + 25b^2$ 10. _____

11. $4x^2 - 24x + 36$ 11. _____

Factor by using either the difference-of-two-squares or the perfect-square trinomial formula.

12. $50x^2 - 60x + 18$ 12. _____

13. $x^4 - 100$ 13. _____

14. $16y^4 - 1$ 14. _____

15. $9x^4 - 30x^2 + 25$ 15. _____

Name _____   Date _____

Practice Set 6.6
A Brief Review of Factoring

Factor, if possible. Be sure to factor completely. Always factor out the greatest common factor first, if one exits.

1. $x^2 + 12x + 27$ 1. _____

2. $y^2 + 20y + 100$ 2. _____

3. $x^2 + 9$ 3. _____

4. $5x^2 + 15x - 90$ 4. _____

5. $25x^2 - 4y^2$ 5. _____

6. $63x - 7x^3$ 6. _____

7. $2x^3 + 8x^2 - 90x$ 7. _____

Name _____ Date _____

8. $3x^2 + 3x + 2$ 8. _____

9. $-3x^3y^2 - 12x^2y^2 - 12xy^2$ 9. _____

10. $2x^2y + 6xy - 72y$ 10. _____

11. $-x^3 + 12x^2 + 45x$ 11. _____

12. $6x^2 + 36x - 2xy - 12y$ 12. _____

13. $2x^2 + 3x + 4$ 13. _____

14. $x^2 + 2x + 3x + 6$ 14. _____

15. $x^4 - 1$ 15. _____

Name _____   Date _____

Practice Set 6.7
Solving Quadratic Equations by Factoring

Using the factoring method, solve for the roots of each quadratic equation. Be sure to place the equation in standard form before factoring. Check your answers.

1. $x^2 - 6x - 16 = 0$ 1. _____

2. $x^2 + 7x - 30 = 0$ 2. _____

3. $2x^2 - 5x - 3 = 0$ 3. _____

4. $x^2 + 5x = 0$ 4. _____

5. $3x^2 - 4x = 0$ 5. _____

6. $3x^2 = 27$ 6. _____

7. $5x^2 - 13x = 2x$ 7. _____

Name _____   Date _____

8. $2x^2 = 5x + 12$

8. _____

9. $x^2 - 35 = 2x$

9. _____

10. $8x^2 = 12 - 10x$

10. _____

11. $9x^2 - 6x = -1$

11. _____

12. $x^2 + 5x = 6$

12. _____

13. $x^2 - 7x = -12$

13. _____

14. $\dfrac{x^2 + x}{2} = 28$

14. _____

15. $\dfrac{x^2 + 5}{2} = 7$

15. _____

Name _____   Date _____

Practice Set 7.1
Simplifying Rational Expressions

Simplify.

1. $\dfrac{7}{x^2-y^2} + \dfrac{2y}{xy^2+y^3}$

   1. _____

2. $\dfrac{3x+12}{x^2+4x}$

   2. _____

3. $\dfrac{2x^2-3x-2}{x^2-2x}$

   3. _____

4. $\dfrac{9x^2-24x+16}{9x^2-16}$

   4. _____

5. $\dfrac{2x^2-7x-15}{x^3-25x}$

   5. _____

Name _____   Date _____

6. $\dfrac{2x^2 - 2x - 12}{4x^2 - 40x + 84}$ 6. _____

7. $\dfrac{4 - 11y - 3y^2}{3y^2 - 7y + 2}$ 7. _____

8. $\dfrac{81 - x^2}{3x^2 - 21x - 54}$ 8. _____

9. $\dfrac{2a^2 - ab - 15b^2}{4a^2 - 25b^2}$ 9. _____

10. $\dfrac{25x^2 - 20xy + 4y^2}{15x^2 - xy - 2y^2}$ 10. _____

Name _____   Date _____

Practice Set 7.2
Multiplying and Dividing Rational Expressions

Multiply.

1. $\dfrac{3x-9}{x-2} \cdot \dfrac{x^2-x-2}{x^2-4x+3}$

    1. _____

2. $\dfrac{32x^3}{8x^2-8} \cdot \dfrac{4x-4}{16x^2}$

    2. _____

3. $\dfrac{x^2-5x-14}{x^2-4x-21} \cdot \dfrac{x^2+7x+12}{x^2+2x-8}$

    3. _____

4. $\dfrac{3x^2-3x}{2x^2-x-1} \cdot \dfrac{2x^2+x-6}{3x^2+6x}$

    4. _____

5. $\dfrac{2x^2-3x}{4x^3+4x^2-9x-9} \cdot \dfrac{2x^2+5x+3}{5x+7}$

    5. _____

93

Name _____   Date _____

Divide.

6. $(4x+7) \div \dfrac{8x^2+10x-7}{2x^2+7x-4}$

6. _____

7. $\dfrac{x+9}{x-2} \div \dfrac{x^2-5x-6}{x^2-8x+12}$

7. _____

8. $\dfrac{9x^2-49}{3x^2+8x-35} \div (3x^2+x-14)$

8. _____

9. $\dfrac{4x^2-2xy-6y^2}{2x+4y} \div \dfrac{4x+4y}{3x+6y}$

9. _____

10. $\dfrac{9x^2-16}{9x^2+24x+16} \div \dfrac{3x^2-x-4}{5x^2-5}$

10. _____

Name _____   Date _____

Practice Set 7.3
Adding and Subtracting Rational Expressions

Find the LCD. Do not combine fractions.

1. $\dfrac{3}{4x+12}, \dfrac{8}{7x+21}$ 1. _____

2. $\dfrac{9}{4x^2y^3z}, \dfrac{17}{3x^5yz^3}$ 2. _____

3. $\dfrac{7}{2x^2-11x+12}, \dfrac{13}{2x^2+7x-15}$ 3. _____

Perform the operation indicated. Be sure to simplify.

4. $\dfrac{x}{3x+1}+\dfrac{x+3}{3x+1}$ 4. _____

5. $\dfrac{2x+1}{7x-1}+\dfrac{3x-4}{7x-1}$ 5. _____

6. $\dfrac{5x}{x+1}-\dfrac{2x-5}{x+1}$ 6. _____

7. $\dfrac{10x-7}{3x+8}-\dfrac{5x-10}{3x+8}$ 7. _____

8. $\dfrac{10x}{(2x-5)(4x+7)}-\dfrac{3x}{(7+4x)(2x-5)}$ 8. _____

Name _____   Date _____

9. $\dfrac{2x+1}{x-1}+\dfrac{x-5}{3x-3}$ 9. _____

10. $\dfrac{3a}{a^2-b^2}+\dfrac{8}{a+b}$ 10. _____

11. $\dfrac{7}{x^2-y^2}+\dfrac{2y}{xy^2+y^3}$ 11. _____

12. $\dfrac{3x-5}{5x-10}-\dfrac{x+3}{x-2}$ 12. _____

13. $\dfrac{6x}{x^2-49}-\dfrac{3}{x+7}$ 13. _____

14. $\dfrac{1}{x^2+3x+2}-\dfrac{2}{x^2-x-6}$ 14. _____

Name _____   Date _____

Practice Set 7.4
Simplifying Complex Rational Expressions

Simplify.

1. $\dfrac{\dfrac{2}{x}}{\dfrac{3}{x}+\dfrac{1}{x^2}}$

1. _____

2. $\dfrac{\dfrac{4}{a}+\dfrac{5}{b}}{\dfrac{2}{ab}}$

2. _____

3. $\dfrac{\dfrac{2}{x}+\dfrac{2}{y}}{x+y}$

3. _____

4. $\dfrac{9-\dfrac{1}{x^2}}{3+\dfrac{1}{x}}$

4. _____

Name _____   Date _____

5. $\dfrac{\dfrac{2x}{x^2-25}}{\dfrac{4}{x+5}-\dfrac{3}{x-5}}$ 5. _____

6. $\dfrac{\dfrac{12}{x^2-36}}{\dfrac{11}{x-6}-\dfrac{3}{x+6}}$ 6. _____

7. $\dfrac{\dfrac{4}{2x+y}-\dfrac{3}{2x-y}}{\dfrac{7}{4x^2-y^2}}$ 7. _____

8. $\dfrac{\dfrac{3}{4x}-\dfrac{8}{3y}}{\dfrac{7}{a}+\dfrac{5}{b}}$ 8. _____

Name _____   Date _____

Practice Set 7.5
Solving Equations Involving Rational Expressions

Solve and check. If there is no solution, say so.

1. $\dfrac{6}{x} + \dfrac{3}{5} = \dfrac{9}{x}$

    1. _____

2. $\dfrac{2}{3x} + \dfrac{1}{6} = \dfrac{6}{x}$

    2. _____

3. $\dfrac{2}{2x+3} = \dfrac{8}{2x-1}$

    3. _____

4. $\dfrac{4}{a^2-1} = \dfrac{2}{a+1} + \dfrac{2}{a-1}$

    4. _____

5. $\dfrac{2m}{m^2-4} + \dfrac{1}{m-2} = \dfrac{2}{m+2}$

    5. _____

Name _____   Date _____

6. $\dfrac{8r}{4r^2-1}=\dfrac{3}{2r+1}+\dfrac{3}{2r-1}$ 6. _____

7. $\dfrac{7}{3a+11}-\dfrac{2}{a-5}=\dfrac{-52}{3a^2-4a-55}$ 7. _____

8. $\dfrac{6}{5a+10}-\dfrac{1}{a-5}=\dfrac{4}{a^2-3a-10}$ 8. _____

9. $\dfrac{x-2}{x^2-4x-5}+\dfrac{x+5}{x^2-25}=\dfrac{2x+13}{x^2+6x+5}$ 9. _____

10. $\dfrac{6z+4}{z-3}=\dfrac{132}{z^2-9}+\dfrac{6z-4}{z+3}$ 10. _____

Name _____   Date _____

Practice Set 7.6
Ratio, Proportion, and Other Applied Problems

Solve.

1. $\dfrac{7}{2} = \dfrac{x}{10}$ 1. _____

2. $\dfrac{7}{12} = \dfrac{y}{8}$ 2. _____

3. $\dfrac{14}{a} = \dfrac{3}{5}$ 3. _____

4. $\dfrac{8}{5} = \dfrac{x}{12}$ 4. _____

5. $\dfrac{3}{10} = \dfrac{11.4}{x}$ 5. _____

6. $\dfrac{6}{7} = \dfrac{x}{12}$ 6. _____

Use a proportion to answer.

7. The scale on a map of Massachusetts is approximately $\dfrac{3}{4}$ inch to 25 miles. If the distance from a college to Boston measures 4 inches on the map, how far apart are the two locations?   7. _____

8. Amy took 5.5 hours to drive 330 miles on the highway. If she continues to drive at the same rate, how many hours will it take her to drive a distance of 510 miles?   8. _____

Name _____   Date _____

9. Alicia is 4 feet tall and casts a shadow that is 9 feet long. At the   9. _____
    same time of day, a tree casts a shadow that is 36 feet long. How
    tall is the tree?

10. A task takes Thomas 4 hours to do by hand, while a machine can   10. _____
    do the task in 1 hour. If Thomas and the machine work on the
    task at the same time, how long will the job take?

11. A family recipe calls for $\frac{2}{3}$ cup of flour and makes 12 servings.   11. _____
    How many cups of flour would you need to make 27 servings?

12. A ramp is 14 meters long and rises 5 meters. A second ramp with   12. _____
    the same angle of incline as the first rises 2.5 meters. How long
    is the second ramp?

13. Two bullet trains travel at speeds of 250 kilometers per hour and   13. _____
    300 kilometers per hour, respectively. If both trains left the
    station at the same time, and the first train has traveled 375
    kilometers, how far has the second train traveled?

14. Toshiro and Mathilda work in the library. Working alone, it takes   14. _____
    Toshiro 7 hours to arrange and stack 1000 books, while it takes
    Mathilda 6.5 hours. To the nearest minute, how long will the job
    take if they work together to arrange and stack 1000 books?

Name _____   Date _____

Practice Set 8.1
Rational Exponents

Simplify and express your answer with positive exponents. Evaluate or simplify the numerical expressions.

1. $\left(\dfrac{2x^{-2}y^3z^{-2}}{3y^4}\right)^2$

   1. _____

2. $\left(\dfrac{2a^2}{b^3}\right)^{-4}$

   2. _____

3. $\left(z^4\right)^{9/8}$

   3. _____

4. $\left(x^{4/5}\right)^4$

   4. _____

5. $\dfrac{a^3}{a^{2/3}}$

   5. _____

6. $x^{5/2} \cdot x^{3/2}$

   6. _____

7. $y^{7/8} \cdot y^{-3/8}$

   7. _____

8. $y^{-1/6}$

   8. _____

Name _____ Date _____

9. $7^{-1/4}$ 9. _____

10. $(-125)^{2/3}$ 10. _____

11. $\left(a^{5/4}b^{2/3}\right)\left(a^{-1/4}b^{2/3}\right)$ 11. _____

12. $\left(\dfrac{81xy^5}{x^4 y^{-1}}\right)^{1/2}$ 12. _____

13. $-3a^{2/3}\left(ab+4a^{4/3}\right)$ 13. _____

Write each expression as one fraction with positive exponents.

14. $y^{-2/3}+3y^{1/3}$ 14. _____

15. $7^{-1/3}+x^{1/2}$ 15. _____

Factor out the common factor $5x$.

16. $10x^{7/4}+25x^{9/8}$ 16. _____

Name _____   Date _____

Practice Set 8.2
Radical Expressions and Functions

Evaluate if possible.

1. $\sqrt{81}$

   1. _____

2. $\sqrt{0.04} + \sqrt{0.25}$

   2. _____

3. $\sqrt{-100}$

   3. _____

4. $-\sqrt[7]{128}$

   4. _____

5. $\sqrt[15]{(7)^{15}}$

   5. _____

6. $\sqrt[3]{-\dfrac{27}{125}}$

   6. _____

Graph each of the following functions. Plot at least four points for each function.

7. $f(x) = \sqrt{x+2}$

8. $g(x) = \sqrt{2x+8}$

105

Name _____   Date _____

For exercises 9–11, assume that variables represent positive real numbers. Replace all radicals with rational exponents.

9. $\sqrt[3]{y^2}$

9. _____

10. $\sqrt[8]{(4m-3n)^5}$

10. _____

11. $\sqrt[6]{\sqrt{a}}$

11. _____

Simplify.

12. $\sqrt{49a^6b^{16}}$

12. _____

13. $\sqrt[4]{81x^{20}y^8}$

13. _____

Change to radical form.

14. $a^{2/5}$

14. _____

15. $4^{-4/7}$

15. _____

16. $(2x+5y)^{5/9}$

16. _____

Name _____   Date _____

Practice Set 8.3
Simplifying, Adding, and Subtracting Radicals

Simplify. Assume that all variables are nonnegative real numbers.

1. $\sqrt{50}$

    1. _____

2. $\sqrt{180}$

    2. _____

3. $\sqrt{25x^3}$

    3. _____

4. $-\sqrt{72k^9 q^6}$

    4. _____

5. $\sqrt[3]{27}$

    5. _____

6. $\sqrt[3]{128x^4}$

    6. _____

7. $\sqrt[3]{-135x^6 y^5 z^8}$

    7. _____

8. $\sqrt[4]{625ab^{19}}$

    8. _____

Name _____ Date _____

Combine.

9. $2\sqrt{3} + 5\sqrt{3}$

9. _____

10. $6\sqrt{7} - 4\sqrt{5} + 4\sqrt{7}$

10. _____

11. $6\sqrt{32} + 5\sqrt{128}$

11. _____

12. $2\sqrt{10} - 4\sqrt{40} + \sqrt{360}$

12. _____

Combine. Assume that all variables represent nonnegative real numbers.

13. $10\sqrt{4x} - 6\sqrt{16x}$

13. _____

14. $6\sqrt{48x^2} - 2\sqrt{27x^2} - \sqrt{3x^2}$

14. _____

15. $5xy^3 \sqrt[3]{8x^5} + 8y^2 \sqrt[3]{64x^8 y^3}$

15. _____

Name _____   Date _____

Practice Set 8.4
Multiplying and Dividing Radicals

Multiply and simplify. Assume that all variables represent nonnegative numbers.

1. $\sqrt{2}\sqrt{5}$

1. _____

2. $\left(2\sqrt{3}\right)\left(-4\sqrt{2}\right)$

2. _____

3. $\left(3a\sqrt{a}\right)\left(5\sqrt{b}\right)$

3. _____

4. $\left(2x\sqrt{5}\right)\left(4\sqrt{5x^2 y}\right)$

4. _____

5. $2\sqrt{x}\left(4\sqrt{3}-7\sqrt{x}\right)$

5. _____

6. $\left(\sqrt{7}+4\right)\left(\sqrt{5}-3\right)$

6. _____

7. $\left(5\sqrt{7}+3\sqrt{10}\right)^2$

7. _____

8. $\left(\sqrt[3]{3}+\sqrt[3]{25}\right)\left(\sqrt[3]{5}-\sqrt[3]{9}\right)$

8. _____

Name _____   Date _____

Divide and simplify. Assume that all variables represent positive numbers.

9. $\sqrt{\dfrac{64}{36}}$

9. _____

10. $\sqrt{\dfrac{18x}{25y^4}}$

10. _____

11. $\dfrac{\sqrt[3]{108x^8 y^7}}{\sqrt[3]{4x^2 y}}$

11. _____

Simplify by rationalizing the denominator.

12. $\dfrac{6}{\sqrt{3}}$

12. _____

13. $\dfrac{2\sqrt{y}}{\sqrt{5x}}$

13. _____

14. $\dfrac{2x}{\sqrt{5}-\sqrt{3}}$

14. _____

15. $\dfrac{\sqrt{6a}-\sqrt{b}}{\sqrt{2a}+\sqrt{2b}}$

15. _____

Name _____   Date _____

Practice Set 8.5
Radical Equations

Solve each radical equation. Check your solution(s).

1. $\sqrt{x+1} = 7$ 1. _____

2. $\sqrt{7a-6} = 6$ 2. _____

3. $\sqrt{10x-5} = 5$ 3. _____

4. $\sqrt{2y+1} - 19 = 0$ 4. _____

5. $\sqrt{x+4} + 8 = 3$ 5. _____

6. $\sqrt{6x^2 + 6x} = 2x$ 6. _____

7. $\sqrt{x+4} = x + 2$ 7. _____

8. $\sqrt{6y+4} + 6 = y$ 8. _____

Name _____   Date _____

9. $\sqrt[3]{3x-2} = 4$ 9. _____

10. $\sqrt[3]{2-6x} + 4 = 6$ 10. _____

Solve each radical equation. This will usually involve squaring each side twice. Check your solutions.

11. $\sqrt{x+8} = \sqrt{x+3} + 1$ 11. _____

12. $\sqrt{x-9} = \sqrt{x+12} - 3$ 12. _____

13. $\sqrt{3x+4} = 2 + \sqrt{x}$ 13. _____

14. $\sqrt{5x+1} - 1 = \sqrt{4-5x}$ 14. _____

15. $\sqrt{2x+6} - \sqrt{x+4} = \sqrt{x-4}$ 15. _____

Name _____   Date _____

Practice Set 8.6
Complex Numbers

Simplify. Express in terms of i.

1. $\sqrt{-225}$ 1. _____

2. $\sqrt{-250}$ 2. _____

3. $3 + \sqrt{-2}$ 3. _____

4. $\dfrac{2}{3} + \sqrt{-18}$ 4. _____

5. $(\sqrt{-24})(\sqrt{-3})$ 5. _____

6. $(\sqrt{-36})(\sqrt{-1})$ 6. _____

Multiply and simplify your answers. Place in i notation before doing any other operations.

7. $(3i)(5i)$ 7. _____

8. $(3 + 2i)(2 + i)$ 8. _____

113

Name _____   Date _____

9. $15i - 7(2i + 3)$ 9. _____

10. $(i\sqrt{5})(i\sqrt{15})$ 10. _____

11. $(6 + \sqrt{-2})(3 - \sqrt{-2})$ 11. _____

Evaluate.

12. i^{36} 12. _____

13. $i^{82} - i^{31}$ 13. _____

Divide.

14. $\dfrac{4-i}{3+i}$ 14. _____

15. $\dfrac{4+3i}{2i}$ 15. _____

16. $\dfrac{6}{5-2i}$ 16. _____

Name _____   Date _____

Practice Set 8.7
Variation

Round all answers to the nearest tenth unless otherwise directed.

1. If y varies directly with x and $y = 15$ when $x = 24$, find y when $x = 36$.

    1. _____

2. If y varies directly with x and $y = 0.5$ when $x = 42$, find y when $x = 193.2$.

    2. _____

3. The distance a spring stretches varies directly with the weight of the object hung on the spring. If a 12-pound weight stretches a spring 15 inches, how far will a 32-pound weight stretch this spring?

    3. _____

4. A car's stopping distance varies directly with the square of its speed. A car that is traveling 20 miles per hour can stop in 20 feet. What distance will it take to stop if it is traveling 50 miles per hour?

    4. _____

5. When an object is dropped, the distance it falls in feet varies directly with the square of the duration of the fall in seconds. A rock dropped from the top of a building falls 16 feet in 1 second. How far will it fall in 3.5 seconds?

    5. _____

6. If y varies inversely with x, and $y = 10$ when $x = 5$, find y when $x = 0.25$.

    6. _____

7. If y varies inversely with the cube of x, and $y = 50$ when $x = 3$, find y when $x = 7$.

    7. _____

Name _____   Date _____

8. The speed of a car varies inversely with the amount of time it takes to cover a certain distance. At 50 mph, a car travels a certain distance in 16 seconds. What is the speed of a car that travels the same distance in 10 seconds?

8. _____

9. If the voltage in an electric circuit is kept at the same level, the current varies inversely with the resistance. The current measures 100 amperes when the resistance is 2.5 ohms. Find the current when the resistance is 20 ohms.

9. _____

10. The amount of light from a light source varies inversely with the square of the distance to the light source. If an object receives 25 lumens when the light source is 10 meters away, how much light will the object receive if the light source is 2 meters away?

10. _____

11. y varies directly with x and inversely with z. $y = 40$ when $x = 6$ and $z = 15$. Find the value of y when $x = 0.5$ and $z = 12$.

11. _____

12. a varies directly with b and inversely with the square root of c. $a = 0.4$ when $b = 2$ and $c = 0.25$. Find the value of a when $b = 15$ and $c = 9$.

12. _____

13. q varies directly with r and the square root of s and inversely with the square of t. $q = 12.5$ when $r = 3$, $s = 36$, and $t = 6$. Find the value of q when $r = 5$, $s = 100$, and $t = 8$.

13. _____

14. The strength of a rectangular beam varies jointly with its width and the square of its thickness. If a beam 6 inches wide and 2 inches thick supports 480 pounds, how much can a beam of the same material that is 8 inches wide and 4.5 inches thick support?

14. _____

15. The force of the wind on a blade of a wind generator varies jointly with the blade's area and the square of the wind velocity. The force of the wind is 72 pounds when the area is 5 square feet and the velocity is 40 feet per second. Find the force when the area is 8 square feet and the velocity is 25 feet per second.

15. _____

Name _____ Date _____

Practice Set 9.1
Quadratic Equations

Solve the equations by using the square root property. Express any complex numbers using *i* notation.

1. $x^2 = 16$

1. _____

2. $2x^2 - 18 = 0$

2. _____

3. $3x^2 - 135 = 0$

3. _____

4. $x^2 + 64 = 0$

4. _____

5. $(x + 16)^2 = 49$

5. _____

6. $(x + 3)^2 = 25$

6. _____

7. $(3x + 1)^2 = 15$

7. _____

8. $4x^2 - 7 = 0$

8. _____

Name _____ Date _____

Solve the equations by completing the square. Express any complex numbers using *i* notation.

9. $x^2 + 6x + 5 = 0$ 9. _____

10. $x^2 + 12x + 35 = 0$ 10. _____

11. $x^2 - 6x + 30 = 0$ 11. _____

12. $x^2 = 16x - 5$ 12. _____

13. $\dfrac{x^2}{2} + \dfrac{7x}{2} = 3$ 13. _____

14. $2x^2 + 2x + 3 = 0$ 14. _____

15. $4x^2 + 3 = x$ 15. _____

16. $3x^2 + 6 = 2x$ 16. _____

Name _____   Date _____

Practice Set 9.2
The Quadratic Formula and Solutions to Quadratic Equations

Solve by the quadratic formula. Simplify your answers.

1. $x^2 + 6x - 6 = 0$

1. _____

2. $2x^2 - 3x - 5 = 0$

2. _____

3. $3x^2 + 12x + 7 = 0$

3. _____

4. $x^2 + x + 9 = 0$

4. _____

5. $4x^2 + 5 = 12$

5. _____

Simplify each equation. Then solve by the quadratic formula. Simplify your answers. Use i notation for nonreal complex numbers.

6. $3x(x - 3) + 2 = 0$

6. _____

7. $5x(x - 2) + 4 = 7(x^2 - 1)$

7. _____

8. $\dfrac{1}{x} - \dfrac{2}{x+2} = \dfrac{1}{5}$

8. _____

Name _____   Date _____

9. $\dfrac{1}{3} - \dfrac{3}{x} = \dfrac{-4}{x-2}$ 9. _____

10. $7x^2 = -3$ 10. _____

11. $5x^2 - 6x + 8 = 0$ 11. _____

Use the discriminant to find what type of solutions (two rational, two irrational, one rational, or two nonreal complex) each of the following equations has. Do not solve the equation.

12. $3x^2 + 6x + 2 = 0$ 12. _____

13. $x^2 - 3(2x - 3) = 0$ 13. _____

14. $5x^2 = 3(x - 1)$ 14. _____

Write a quadratic equation having the given solutions.

15. -3 and $-\dfrac{1}{2}$ 15. _____

16. $i\sqrt{5}$ and $-i\sqrt{5}$ 16. _____

Name _____   Date _____

Practice Set 9.3
Equations That Can Be Transformed into Quadratic Form

Solve. Express any nonreal complex numbers with *i* notation.

1. $x^4 - 13x^2 + 40 = 0$
1. _____

2. $x^4 + 10x^2 + 21 = 0$
2. _____

Solve for real roots.

3. $6x^8 - 7x^4 + 1 = 0$
3. _____

4. $x^6 + 27x^3 = 0$
4. _____

5. $x^6 = 2x^3 + 8$
5. _____

6. $x^8 - 16x^4 + 15 = 0$
6. _____

7. $x^6 - 4x^3 = 12$
7. _____

Name _____ Date _____

8. $x^{2/3} - 9x^{1/3} + 20 = 0$ 8. _____

9. $6x^{2/5} + 18x^{1/5} + 12 = 0$ 9. _____

10. $x^{2/3} + 8x^{1/3} + 12 = 0$ 10. _____

11. $(4x - 7)^2 - 4(4x - 7) = 12$ 11. _____

12. $(x^2 + 2x)^2 - 14(x^2 + 2x) - 15 = 0$ 12. _____

13. $x - 7x^{1/2} - 18 = 0$ 13. _____

14. $3x^{-2} + 3x^{-1} = 168$ 14. _____

15. $5x^{-2} - 2x^{-1} = 0$ 15. _____

Name _____   Date _____

Practice Set 9.4
Formulas and Applications

Solve for the variable specified. Assume that all other variables are nonzero.

1. $A = 6s^2$; for s

    1. _____

2. $s = \dfrac{1}{2}gt^2$; for t

    2. _____

3. $V = \dfrac{1}{3}\pi r^2 h$; for r

    3. _____

4. $6a^2 - 5 = 8m$; for a

    4. _____

5. $J = \dfrac{2mnW^2}{3p}$; for W

    5. _____

6. $5ay^2 + 6by = 0$; for y

    6. _____

7. $8x^2 + yx + 5 = 0$; for x

    7. _____

8. $(a + 2)x^2 - 7x + 3y = 0$; for x

    8. _____

Name _____   Date _____

Use the Pythagorean theorem to find the missing side(s).

9. $c = \sqrt{17}$, $a = 2\sqrt{2}$; find b 9. _____

10. $c = 10$, $b = 3a$; find a and b 10. _____

Solve.

11. A lawn has the shape of a right triangle. Its hypotenuse is 15 meters long and the two legs are equal in length. How long are the legs of the triangle?   11. _____

12. A rectangular piece of cardboard has a length that is 0.2 feet greater than its width. The area of the cardboard is 0.63 square feet. Find the length and width of the cardboard.   12. _____

13. The area of a triangular flag is 35 square inches. Its altitude is 4 inches longer than twice its base. Find the lengths of the altitude and the base.   13. _____

14. The area of the wall of a house is 390 square feet. Its width is 9 feet less than 3 times its height. Find the height and width of the wall of the house.   14. _____

15. Andrew drove at a constant speed on a dirt road for 75 miles. He then traveled 20 mph faster on a paved road for 180 miles. If he drove for 7 hours, find the car's speed for each part of the trip.   15. _____

Name _____   Date _____

Practice Set 9.5
Quadratic Functions

Find the coordinates of the vertex and the intercepts of each of the following quadratic functions. When necessary, approximate x-intercepts to the nearest tenth.

1. $f(x) = x^2 + 6x + 8$ 1. _____

2. $g(x) = x^2 + 5x - 6$ 2. _____

3. $p(x) = 2x^2 + 4x + 10$ 3. _____

4. $h(x) = -x^2 + 9x - 14$ 4. _____

5. $f(x) = 5x^2 + 17x - 12$ 5. _____

6. $f(x) = 3x^2 - 4x - 16$ 6. _____

In each of the following exercises, find the vertex, the y-intercept, and the x-intercepts (if any exist), and then graph the function.

7. $f(x) = x^2 - 3x - 4$ 8. $g(x) = x^2 + 3x - 4$

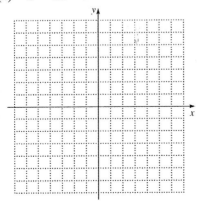
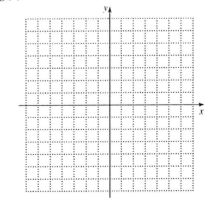

Name _____   Date _____

9. $f(x) = -x^2 - 8x - 12$

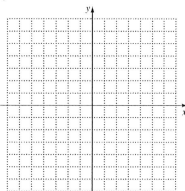

10. $h(x) = -x^2 + 8x - 12$

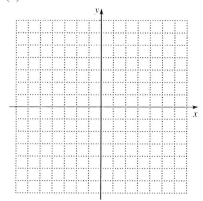

11. $p(x) = 2x^2 - 6x$

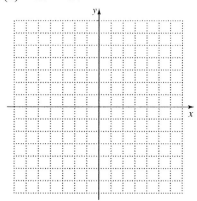

12. $r(x) = 3x^2 - 2x + 1$

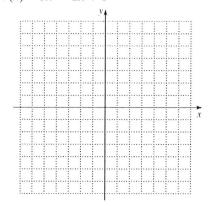

13. $f(x) = -x^2 + 2x - 1$

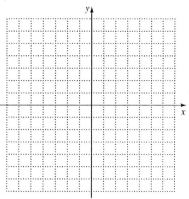

14. $s(x) = -x^2 + 4$

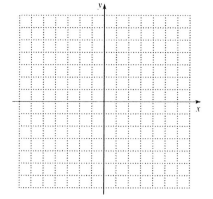

Name _____   Date _____

Practice Set 9.6
Compound Inequalities and Quadratic Inequalities

Graph the values of x that satisfy the conditions given.

1. $1 < x$ and $x < 8$

2. $0 \leq x \leq 5$

3. $-4 < x \leq \dfrac{1}{2}$

4. $x \leq -3$ or $x > 2$

5. $x \geq 3$ or $x \leq -2$

6. $x < 0$ or $x > 7$

Solve for x and graph your results.

7. $4 < x + 1$ and $8x - 3 < 45$ 7. _____

8. $\dfrac{x}{3} + 1 < 1$ or $\dfrac{x}{7} \geq 1$ 8. _____

Name _____ Date _____

Solve and graph.

9. $x^2 - 10x + 16 < 0$ 9. _____   ⟵———————⟶ x

10. $x^2 - 4x \geq 5$ 10. _____   ⟵———————⟶ x

11. $2x^2 - 3x - 5 \leq 0$ 11. _____   ⟵———————⟶ x

Solve.

12. $x^2 + 9x + 8 \leq 0$ 12. _____

13. $10x^2 + 11x < 6$ 13. _____

14. $x^2 - 14x \leq -49$ 14. _____

Solve each of the following quadratic equations if possible. Round your answers to the nearest tenth.

15. $x^2 + 5x > 9$ 15. _____

16. $7x^2 \geq 5x^2 - 8$ 16. _____

Name _____ Date _____

Practice Set 9.7
Absolute Value Equations and Inequalities

Solve each absolute value equation. Check your solutions.

1. $|y| = 23$

2. $|y - 9| = 13$

3. $|4x| = 0.82$

4. $|3m + 2| - 10 = -6$

5. $2\left|3 + \dfrac{1}{2}x\right| + 22 = 28$

6. $|4s + 9| = |s + 2|$

7. $\left|\dfrac{1}{2}x + 2\right| = \left|\dfrac{3}{4}x - 2\right|$

8. $\left|\dfrac{4x + 3}{5}\right| = |3x + 7|$

Name _____   Date _____

Solve and graph the solutions.

9. $|x| < 4$

9. _____   ⟶ x

10. $\left|x + \dfrac{3}{2}\right| > \dfrac{7}{2}$

10. _____   ⟶ x

Solve for x.

11. $|x - 9| < 4$

11. _____

12. $|x + 3| - 4 \leq -2$

12. _____

13. $\left|\dfrac{3}{5}(x - 2)\right| < 6$

13. _____

14. $|8x - 2| > 3$

14. _____

15. $\left|\dfrac{3}{4}x - \dfrac{1}{4}\right| \geq 3$

15. _____

Solve.

16. In a certain company, the measured thickness t of a computer chip must not differ from the standard s by more than 0.05 millimeter. The engineers express this requirement as $|t - s| \leq 0.05$. Find the limits of t if the standard s is 1.33 millimeters.

16. _____

Name _____   Date _____

Practice Set 10.1
The Distance Formula and the Circle

Find the distance between each pair of points. Simplify your answers.

1. $(-2, 5)$ and $(-2, 2)$ 1. _____

2. $(1, 3)$ and $(4, 7)$ 2. _____

3. $(2, -3)$ and $(-1, 3)$ 3. _____

4. $(-0.5, 8.2)$ and $(3.5, 6.2)$ 4. _____

5. $\left(\dfrac{1}{3}, \dfrac{1}{5}\right)$ and $\left(\dfrac{5}{3}, \dfrac{6}{5}\right)$ 5. _____

Write in standard form the equation of the circle with the given center and radius.

6. center $(-2, 3)$; $r = 4$ 6. _____

7. center $(3.2, -1.4)$; $r = \dfrac{3}{7}$ 7. _____

8. center $\left(0, \dfrac{6}{5}\right)$; $r = \sqrt{13}$ 8. _____

Name _____   Date _____

Give the center and radius of each circle. Then sketch its graph.

9. $x^2 + y^2 = 36$

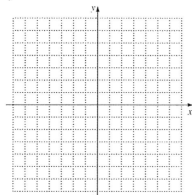

10. $(x - 2)^2 + (y + 3)^2 = 9$

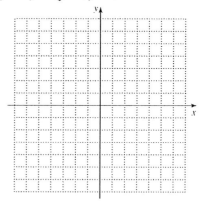

Rewrite each equation in standard form. Find the center and radius of each circle.

11. $x^2 + 2x + y^2 - 4y - 20 = 0$ 11. _____

12. $x^2 - 6x + y^2 - 4y = 3$ 12. _____

13. $x^2 + y^2 + 2y = 3$ 13. _____

14. $x^2 + 10x + y^2 - 6y + 29 = 0$ 14. _____

15. $x^2 + y^2 + 3y + 5x = 0$ 15. _____

Name _____   Date _____

Practice Set 10.2
The Parabola

Graph each parabola and label the vertex. Find the *y*-intercept.

1. $y = 3x^2$

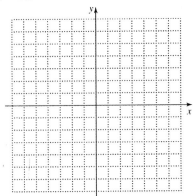

2. $y = -x^2 + 4$

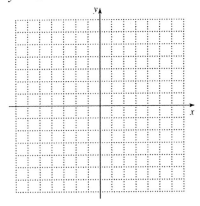

3. $y = (x-1)^2 + 4$

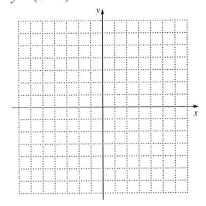

4. $y = 2(x+4)^2 - 3$

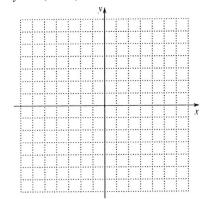

5. $y = \dfrac{1}{2}(x-3)^2 + \dfrac{3}{2}$

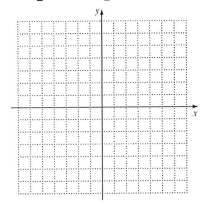

6. $y = -2\left(x - \dfrac{3}{2}\right)^2 + 3$

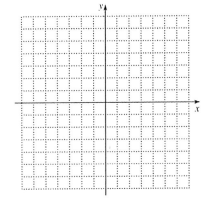

Name _____ Date _____

Graph each parabola and label the vertex. Find the *x*-intercept.

7. $x = \dfrac{1}{4}y^2$

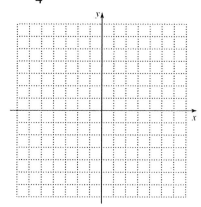

8. $x = -\dfrac{1}{2}y^2 + 3$

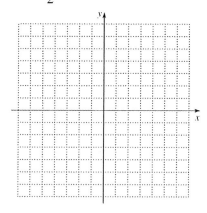

9. $x = (y-2)^2 - 2$

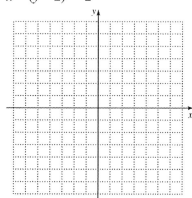

10. $x = 3(y+1)^2 - 3$

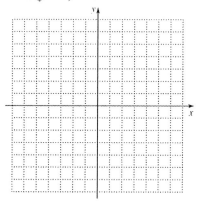

Rewrite each equation in standard form. Determine (a) whether the parabola is horizontal or vertical, (b) the direction it opens, and (c) the vertex.

11. $y = x^2 - 8x + 10$

11. _____

a. _____

b. _____

c. _____

12. $x = -3y^2 + 12y + 6$

12. _____

a. _____

b. _____

c. _____

134

Name _____ Date _____

Practice Set 10.3
The Ellipse

Graph each ellipse. Label the intercepts.

1. $\dfrac{x^2}{4}+\dfrac{y^2}{36}=1$

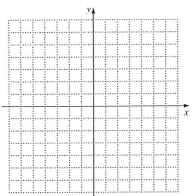

2. $\dfrac{x^2}{16}+\dfrac{y^2}{9}=1$

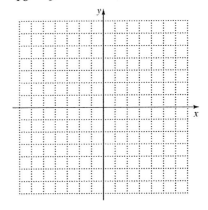

3. $\dfrac{x^2}{25}+\dfrac{y^2}{1}=1$

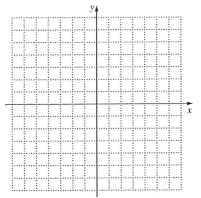

4. $\dfrac{x^2}{9}+\dfrac{y^2}{25}=1$

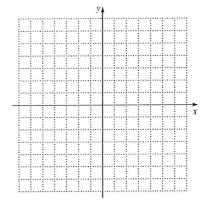

5. $x^2+4y^2=16$

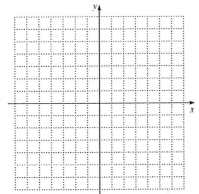

6. $4x^2+5y^2=80$

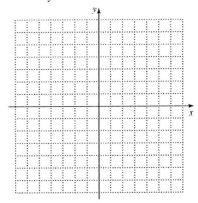

135

Name _____ Date _____

7. $\dfrac{x^2}{\frac{16}{9}} + \dfrac{y^2}{\frac{121}{4}} = 1$

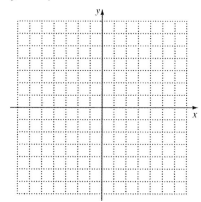

8. $\dfrac{x^2}{\frac{169}{9}} + \dfrac{y^2}{\frac{25}{4}} = 1$

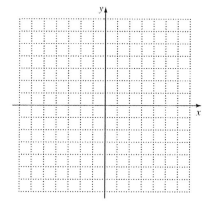

Graph each ellipse. Label the center.

9. $\dfrac{(x+2)^2}{16} + \dfrac{(y-3)^2}{9} = 1$

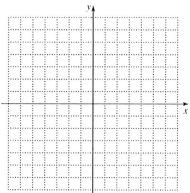

10. $\dfrac{(x-1)^2}{25} + \dfrac{(y+1)^2}{1} = 1$

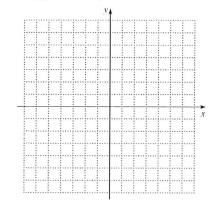

11. $\dfrac{(x-2)^2}{9} + \dfrac{y^2}{16} = 1$

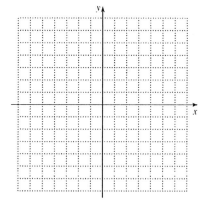

12. $\dfrac{x^2}{16} + \dfrac{(y+4)^2}{1} = 1$

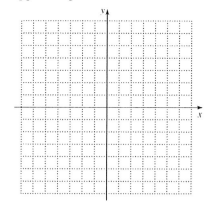

Practice Set 10.4
The Hyperbola

Find the vertices and graph each hyperbola.

1. $\dfrac{x^2}{9} - \dfrac{y^2}{4} = 1$

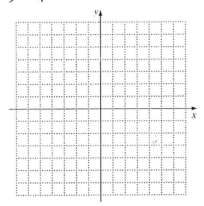

2. $\dfrac{x^2}{4} - \dfrac{y^2}{36} = 1$

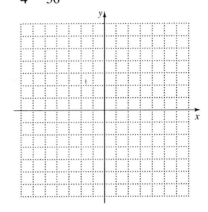

3. $\dfrac{y^2}{4} - \dfrac{x^2}{1} = 1$

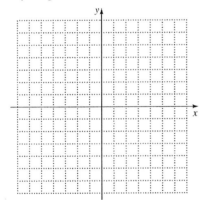

4. $\dfrac{y^2}{25} - \dfrac{x^2}{9} = 1$

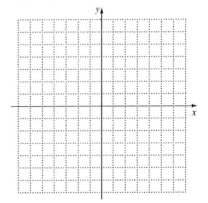

5. $9x^2 - y^2 = 9$

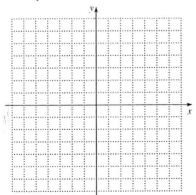

6. $y^2 - x^2 = 16$

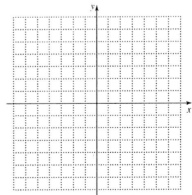

Name _____   Date _____

7. $8y^2 - 4x^2 = 64$

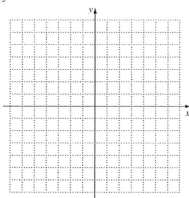

8. $5x^2 - 12y^2 = 60$

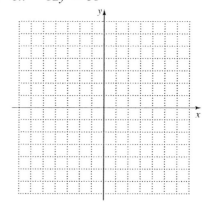

Find the center and then graph each hyperbola.

9. $\dfrac{(x+2)^2}{4} - \dfrac{(y+1)^2}{9} = 1$

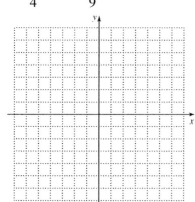

10. $\dfrac{(x-3)^2}{1} - \dfrac{(y+1)^2}{16} = 1$

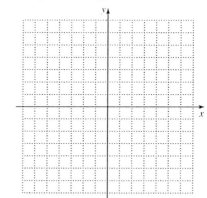

11. $\dfrac{(y+1)^2}{9} - \dfrac{(x-1)^2}{9} = 1$

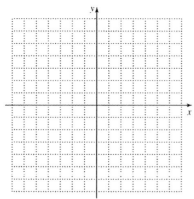

12. $\dfrac{(y-2)^2}{4} - \dfrac{(x-2)^2}{16} = 1$

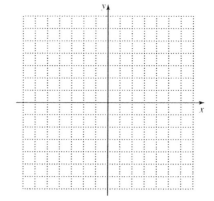

Name _____   Date _____

Practice Set 10.5
Nonlinear Systems of Equations

Solve each of the following systems by the substitution method. Graph each equation to verify that the answer seems reasonable.

1. $y = x^2 - 4$
    $y = x + 2$ _____

2. $16x^2 + 12y^2 = 192$
    $3y + 2x = 0$ _____

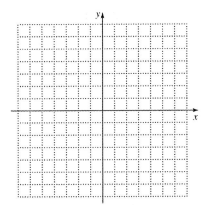

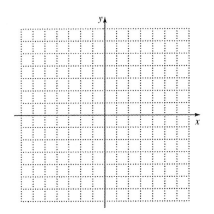

Solve each of the following systems by the substitution method.

3. $x^2 + y^2 = 61$
    $x + y = -11$

    3. _____

4. $x^2 + y^2 = 25$
    $x + y = 7$

    4. _____

5. $y = x^2 - 12x + 36$
    $x + y = 6$

    5. _____

6. $x^2 - 49y^2 - 25 = 0$
    $x + 7y - 2 = 0$

    6. _____

7. $\dfrac{y^2}{16} - \dfrac{x^2}{16} = 1$
    $3y - 5x = 0$

    7. _____

Name _____   Date _____

Solve each of the following systems by the addition method. Graph each equation to verify that the answer seems reasonable.

8. $x^2 + y^2 = 25$
    $20x^2 - 5y^2 = 100$

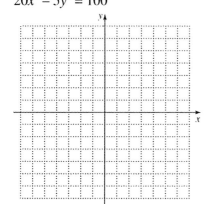

9. $4x^2 + 25y^2 = 100$
    $16y^2 - 4x^2 = 64$

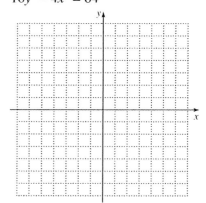

Solve each of the following systems by the addition method.

10. $x^2 + y^2 = 125$
    $x^2 - y^2 = 75$

10. _____

11. $x^2 + y^2 = 36$
    $x^2 - y^2 = 36$

11. _____

12. $4x^2 = 3y^2 + 24$
    $2(x^2 - 15) = -3y^2$

12. _____

13. $x^2 = 1 + 3y^2$
    $2x^2 = 19 - 3y^2$

13. _____

14. $x^2 + 3y^2 = 17$
    $x^2 - 5y^2 = 12$

14. _____

Name _____   Date _____

Practice Set 11.1
Function Notation

For the function $f(x) = 4x - 7$, find the following.

1. $f\left(\dfrac{2}{3}\right)$

    1. _____

2. $f(a + 2)$

    2. _____

3. $f(b^2) - f(12)$

    3. _____

If $g(x) = 5x^2 - 8x + 9$, find the following.

4. $g(-3)$

    4. _____

5. $g(a + 1)$

    5. _____

6. $g\left(-\dfrac{3b}{4}\right)$

    6. _____

If $h(x) = \sqrt{x+2}$, find the following.

7. $h(10)$

    7. _____

8. $h(4a^2 + 6)$

    8. _____

Name _____   Date _____

9. $h(b^2 + 3b - 2)$ 9. _____

If $s(x) = \dfrac{5}{x+2}$, find the following.

10. $s(5)$ 10. _____

11. $s(4a^3 + 1)$ 11. _____

12. $s\left(-\dfrac{4}{3}\right) + s(-3)$ 12. _____

Find $\dfrac{f(x+h) - f(x)}{h}$ for the following functions.

13. $f(x) = 6x - 5$ 13. _____

14. $f(x) = 4x^2$ 14. _____

Solve.

15. The surface area of a cube is given by $S = 6a^2$ where a is the length of a side.
   a. Write the surface area of a cube as a function of the side length a.
   b. Find the surface area of a cube with side length 4 feet.
   c. Suppose that an error is made and the side length is calculated to be $(4 + e)$ feet. Find an expression for the surface area as a function of the error e.
   d. Evaluate the surface area for $a = (4 + e)$ feet when $e = 0.3$. What is the difference in the surface area due to the error in measurement?

15. a._____

b._____

c._____

d._____

Name _____   Date _____

Practice Set 11.2
General Graphing Procedures for Functions

Determine whether or not each graph represents a function.

1. _____

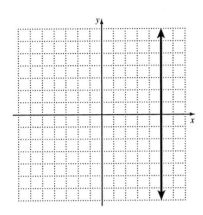

2. _____

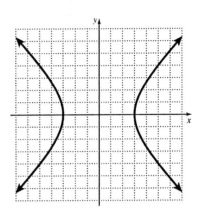

3. _____

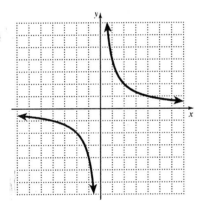

4. _____

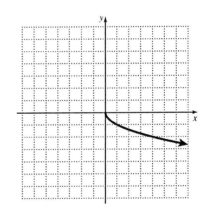

For each of exercises 5–12, graph the two functions on one coordinate plane.

5. $f(x) = x^2$
   $g(x) = x^2 - 2$

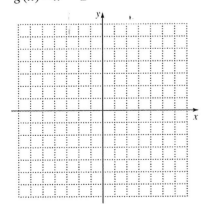

6. $f(x) = x^2$
   $g(x) = (x+3)^2 - 2$

Name _____   Date _____

7. $f(x) = x^3$
   $g(x) = x^3 - 3$

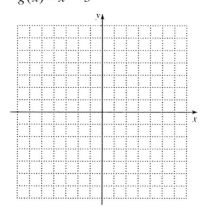

8. $f(x) = x^3$
   $g(x) = (x+2)^3 + 2$

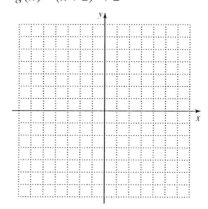

9. $f(x) = |x|$
   $g(x) = |x-4|$

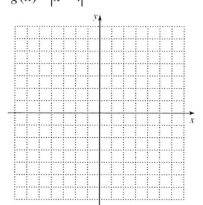

10. $f(x) = |x|$
    $g(x) = |x-2| - 5$

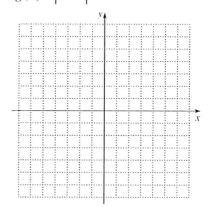

11. $f(x) = \dfrac{4}{x}$
    $g(x) = \dfrac{4}{x} - 3$

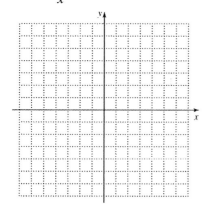

12. $f(x) = \dfrac{3}{x}$
    $g(x) = \dfrac{3}{x-2}$

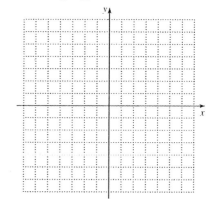

Name _____   Date _____

Practice Set 11.3
Algebraic Operations on Functions

For the following functions, find **(a)** $(f+g)(x)$, **(b)** $(f-g)(x)$, **(c)** $(f+g)(3)$, and **(d)** $(f-g)(-2)$.

1. $f(x) = 3x - 2$
   $g(x) = -4x + 7$

   1. a._____
      b._____
      c._____
      d._____

2. $f(x) = 7 - 5x$
   $g(x) = x^2 - x + 7$

   2. a._____
      b._____
      c._____
      d._____

3. $f(x) = 1.6x^3 - 2.7x$
   $g(x) = 4.6x^2 + 7.6$

   3. a._____
      b._____
      c._____
      d._____

4. $f(x) = 4\sqrt{2x+9}$
   $g(x) = -\sqrt{2x+9}$

   4. a._____
      b._____
      c._____
      d._____

For the following functions, find **(a)** $(fg)(x)$, **(b)** $\left(\dfrac{f}{g}\right)(x)$, **(c)** $(fg)(3)$, and **(d)** $\left(\dfrac{f}{g}\right)(-3)$.

5. $f(x) = x - 3$
   $g(x) = 2x$

   5. a._____
      b._____
      c._____
      d._____

145

Name _____ Date _____

6. $f(x) = x + 2$
    $g(x) = x^2 - 5x - 14$

    6. a. _____
       b. _____
       c. _____
       d. _____

7. $f(x) = x^2 - 8x + 16$
    $g(x) = x - 4$

    7. a. _____
       b. _____
       c. _____
       d. _____

8. $f(x) = \sqrt{x+6}$
    $g(x) = 2x$

    8. a. _____
       b. _____
       c. _____
       d. _____

Find $f[g(x)]$ for each of the following.

9. $f(x) = x - 5$
    $g(x) = 6 - 2x$

    9. _____

10. $f(x) = 11 - 4x$
    $g(x) = x^2 - 2$

    10. _____

11. $f(x) = \dfrac{4}{5x+1}$
    $g(x) = 2x - 2$

    11. _____

12. $f(x) = \left|\dfrac{4}{3}x + 1\right|$
    $g(x) = -3x - 2$

    12. _____

Name _____   Date _____

Practice Set 11.4
Inverse of a Function

Indicate whether each function is one-to-one.

1. $C = \{(1, 4.5), (3, 7.5), (-4, -7.5), (7.5, 4)\}$ 1. _____

2. $A = \{(4, 5), (5, 5), (6, 6)\}$ 2. _____

3. $B = \{(7, 9), (9, 7), (-7, -9), (-9, -7)\}$ 3. _____

Indicate whether each graph represents a one-to-one function.

4. _____   5. _____

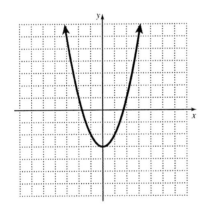 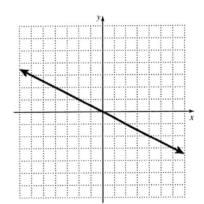

Find the inverse of each function.

6. $f(x) = x - 4$

6. _____

7. $f(x) = 4x + 2$ 7. _____

147

Name _____   Date _____

8. $f(x) = x^3 - 5$ 8. _____

9. $f(x) = \dfrac{4}{x}$ 9. _____

10. $f(x) = \dfrac{5}{3x-4}$ 10. _____

Find the inverse of each function. Graph the function and its inverse on one coordinate plane. Graph the line $y = x$ as a dashed line.

11. $f(x) = 2x + 2$ _____   12. $g(x) = -3x - 4$ _____

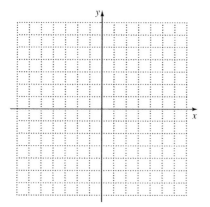
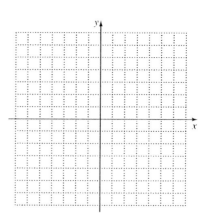

13. $h(x) = \dfrac{1}{2}x + 3$ _____   14. $k(x) = -\dfrac{4}{3}x$ _____

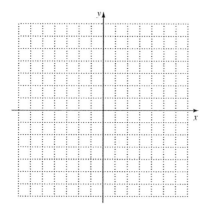
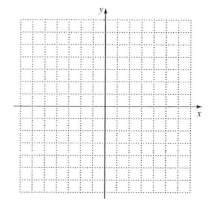

Name _____   Date _____

Practice Set 12.1
The Exponential Function

Graph each function.

1. $f(x) = 4^x$

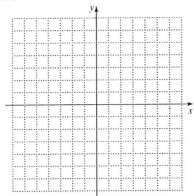

2. $f(x) = 3^{-x}$

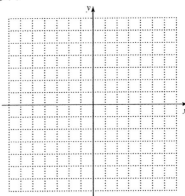

3. $f(x) = 2^{x+4}$

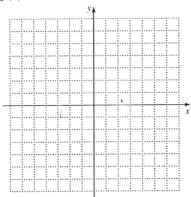

4. $f(x) = 4^{x-1}$

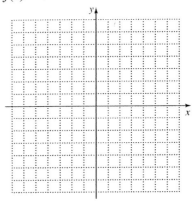

5. $f(x) = 2^x + 3$

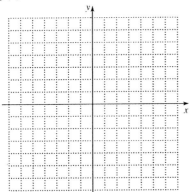

6. $f(x) = 3^x - 4$

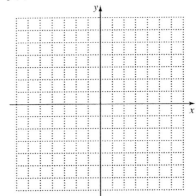

Name _____   Date _____

Solve for x.

7. $6^x = 36$

7. _____

8. $4^x = \dfrac{1}{4}$

8. _____

9. $5^x = 1$

9. _____

10. $2^{-x} = \dfrac{1}{16}$

10. _____

11. $5^{x+3} = 25$

11. _____

12. $7^{2x-5} = 343$

12. _____

To solve exercises 13–14, use the interest formula $A = P\left(1 + \dfrac{r}{n}\right)^{nt}$. Round your answers to the nearest cent.

13. Anton is investing $4000 at an annual rate of 4.6% compounded annually. How much money will Anton have after 5 years?

13. _____

14. How much money will Ayah have in 3 years if she invests $2000 at a 6.6% annual rate of interest compounded quarterly? How much will she have if it is compounded monthly?

14. _____

Name _____   Date _____

Practice Set 12.2
The Logarithmic Function

Write in logarithmic form.

1. $216 = 6^3$

2. $0.0001 = 10^{-4}$

3. $\dfrac{1}{64} = 4^{-3}$

Write in exponential form.

4. $3 = \log_3 27$

5. $\dfrac{1}{2} = \log_{49} 7$

6. $-6 = \log_e x$

Solve.

7. $\log_8 x = -2$

1. _____

2. _____

3. _____

4. _____

5. _____

6. _____

7. _____

Name _____   Date _____

8. $\log_9 x = 2$

8. _____

9. $\log_3\left(\dfrac{1}{81}\right) = x$

9. _____

10. $\log_a 64 = 6$

10. _____

11. $\log_{729} 9 = x$

11. _____

Evaluate.

12. $\log_{10}(0.00001)$

12. _____

13. $\log_{10} \sqrt{10}$

13. _____

Graph.

14. $\log_2 x = y$

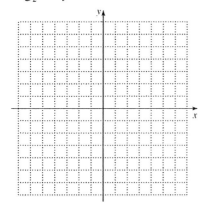

15. $\log_{1/5} x = y$

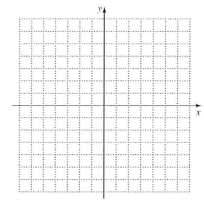

Name _____   Date _____

Practice Set 12.3
Properties of Logarithms

Write each expression as a sum or difference of logarithms x, y, and z.

1. $\log_3 xy$ 1. _____

2. $\log_8 \left(\dfrac{z}{x}\right)$ 2. _____

3. $\log_5 y^3$ 3. _____

4. $\log_7 x^3 y z^2$ 4. _____

5. $\log_2 \left(\dfrac{5y^3 z}{\sqrt{x}}\right)$ 5. _____

6. $\log_3 \left(\sqrt[4]{\dfrac{x^5 y^3}{z}}\right)$ 6. _____

Write as a single logarithm.

7. $\log_2 12 + \log_2 x + \log_2 5$ 7. _____

8. $4 \log_9 x + \log_9 5 - 2 \log_9 y$ 8. _____

Name _____   Date _____

9. $\frac{1}{3}\log_a 6 + 2\log_a 6 - 5\log_a x$ 9. _____

Use the properties of logarithms to simplify each of the following.

10. $\log_5 5$ 10. _____

11. $\frac{1}{5}\log_3 3 - \log_{11} 1$ 11. _____

Find x in each of the following.

12. $\log_2 x = \log_2 12$ 12. _____

13. $\log_6 (3x + 3) = \log_6 36$ 13. _____

14. $\log_4 x + \log_4 2 = 3$ 14. _____

15. $\log_{12} 6 = \log_{12} x - \log_{12} 5$ 15. _____

16. $\log_3 (6x + 12) - \log_3 (x - 5) = 2$ 16. _____

Practice Set 12.4
Common Logarithms, Natural Logarithms, and Change of Base Logarithms

Approximate the following with a scientific calculator or graphing calculator.

1. log 11.2

    1. _____

2. antilog (−1.902)

    2. _____

3. ln 8.22

    3. _____

4. log 0.0045

    4. _____

5. antilog$_e$(7.254)

    5. _____

6. ln 0.00756

    6. _____

Find an approximate value of x using a scientific calculator or a graphing calculator.

7. log x = 1.37

    7. _____

8. log x = −0.153

    8. _____

Name _____   Date _____

9. $\ln x = 3.2$ 9. _____

10. $\ln x = 0.85$ 10. _____

11. $\log x = 3.871$ 11. _____

12. $\ln x = -0.08$ 12. _____

Use a scientific calculator or a graphing calculator and common logarithms to evaluate the following.

13. $\log_2 5.4$ 13. _____

14. $\log_{12} 0.451$ 14. _____

Use a scientific calculator or a graphing calculator and natural logarithms to evaluate the following.

15. $\log_4 0.0332$ 15. _____

16. $\log_6 4.23$ 16. _____

Name _____   Date _____

Practice Set 12.5
Exponential and Logarithmic Equations

Solve each logarithmic equation and check your solutions.

1. $\log 2x + \log 4 = \log(2x + 12)$ 1. _____

2. $\log(6 - x) = \log(x - 4) + \log 5$ 2. _____

3. $\log_7 3 - \log_7 x = \log_7 6$ 3. _____

4. $\log_2(x - 5) + \log_2(x - 11) = 4$ 4. _____

5. $\ln(5x - 1) = \ln 1 - \ln(x - 1)$ 5. _____

6. $\log_{11}(5x - 6) + \log_{11} x = 1$ 6. _____

7. $\log(13 + 4x) = 2 \log(x + 2)$ 7. _____

8. $\ln 8 - \ln x = \ln(x - 7)$ 8. _____

Name _____   Date _____

9. $2\log_6(x-1) = \log_6(x+11)$ 9. _____

Solve each exponential equation. Leave your answers in exact form. Do not approximate.

10. $12^{x+6} = 7$ 10. _____

11. $3^{3x+5} = 15$ 11. _____

Solve each exponential equation. Use your calculator to approximate your solutions to the nearest thousandth.

12. $8^{2x+3} = 135$ 12. _____

13. $6^x = 4^{x+2}$ 13. _____

14. $52 = e^{4x-2}$ 14. _____

When a principal P earns an annual interest rate r compounded yearly, the amount A after t years is $A = P(1 + r)^t$. Use this information to solve exercises 15–16. Round all answers to the nearest whole year.

15. How long will it take for $4000 to grow to $7000 at 6% compounded annually?   15. _____

16. How long will it take for a principal to triple at 4% compounded annually?   16. _____

Practice Set Answers Chapter 0

0.1

1. $\dfrac{2}{3}$
3. $\dfrac{5}{7}$
5. $4\dfrac{2}{3}$
7. $13\dfrac{1}{2}$
9. $\dfrac{7}{3}$
11. $\dfrac{29}{17}$
13. 14
15. 96

0.2

1. 9
3. 60
5. 100
7. $\dfrac{7}{10}$
9. $\dfrac{5}{9}$
11. $3\dfrac{9}{14}$
13. $3\dfrac{7}{36}$
15. $3\dfrac{1}{2}$

0.3

1. $\dfrac{4}{7}$
3. $\dfrac{6}{11}$
5. $\dfrac{2}{49}$
7. $\dfrac{4}{7}$
9. $\dfrac{2}{3}$
11. $\dfrac{11}{2}$ or $5\dfrac{1}{2}$
13. 132 Big Burgers
15. 14 pieces

0.4

1. 5.6; five and six tenths
3. $14\dfrac{27}{10,000}$; fourteen and twenty-seven ten thousandths
5. 81.541
7. 6.1368
9. 0.00075
11. 914.1
13. 1678.5
15. 0.059056

0.5

1. 32%
3. 39.8%
5. 0.576%
7. 0.0793
9. 6.238
11. 50
13. 260%
15. 1400

Practice Set Answers Chapter 1

1.1
1. 1, 2, 4, 5
3. 3, 5
5. $-43; 3.2; -\pi; \frac{7}{8}$
7. -25
9. $-3\frac{1}{2}$
11. -5
13. 9.14
15. $-\frac{7}{36}$

1.2
1. -13
3. 11
5. 0
7. -3.29
9. $\frac{53}{5}$ or $10\frac{3}{5}$
11. 12
13. 0.68
15. $89 - (-15)$; 104 degrees

1.3
1. 224
3. 1560
5. -462.94
7. 70
9. 4
11. 29
13. undefined
15. 7

1.4
1. 4^5
3. x^6
5. 10,000,000
7. $\frac{9}{49}$
9. -125
11. -1372
13. 53
15. -17

1.5
1. 7
3. 1
5. 38
7. 29
9. -10
11. 2
13. $-\frac{31}{5}$ or $-6\frac{1}{5}$
15. 31.36

1.6
1. $3x + 6$
3. $-12x - 6$
5. $-12x - 16y - 8z$
7. $\frac{1}{6}x + \frac{1}{5}y$
9. $\frac{x^2}{3} + \frac{2xy}{3} + \frac{2xz}{3}$
11. $-18x + 12y - 6z$
13. $18(5 + 3x) = 90 + 54x$ square inches
15. $25(3\frac{1}{2} + 8a - 5) = 200a - 37\frac{1}{2}$ square feet

1.7

1. $11x^2$
3. $4a^4 - 11a^2$
5. $3.5x + 0.9y$
7. $x - 4y - 3z$
9. $-\frac{1}{5}x - \frac{1}{3}y$
11. $16x + 12$
13. $13b + 34$
15. $2ab + 13ac - 3bc + 30cd$

1.8

1. -14
3. -3
5. $-\frac{1}{3}$
7. -19
9. 14
11. 122
13. 384 square inches
15. $1584

1.9

1. $2n - 10$
3. $-5a - 9b$
5. $21x$
7. $6x^3 - 9x^2 - 36x$
9. $-90x + 30y$
11. $9b^2 - 2b$
13. $x - 9y + 10$
15. $-10x^2 - 16x + 20$

Practice Set Answers Chapter 2

2.1
1. 6
3. 6
5. 3
7. 18
9. $\dfrac{7}{10}$
11. no; 3
13. yes
15. no; −29

2.2
1. 4
3. 13
5. 22
7. $-\dfrac{5}{3}$ or $-1\dfrac{2}{3}$
9. 7
11. no; 7
13. yes
15. no; −4

2.3
1. 5
3. 4
5. 6
7. 6
9. $-\dfrac{7}{2}$ or $-3\dfrac{1}{2}$
11. 7
13. 3
15. $\dfrac{13}{3}$ or $4\dfrac{1}{3}$

2.4
1. 2
3. 3
5. $-\dfrac{3}{2}$ or $-1\dfrac{1}{2}$
7. 3
9. −2
11. −6
13. $\dfrac{2}{3}$
15. −2.3

2.5
1. $x + 7$
3. $x - 18$
5. $\dfrac{1}{4}x - 1$
7. $3x - 16$
9. $\dfrac{1}{2}(x + 2)$
11. x = value of Allison's car; $x + 2300$ = value of Alicia's car
13. j = cost of John's dinner; $\dfrac{1}{3}j + 4$ = cost of Raymond's dinner
15. s = number of Scott's comic books; $s + 17$ = number of Patricia's comic books; $3s$ = number of Walter's comic books; $4s - 5$ = number of Adrienne's comic books

2.6

1. 546
3. 15
5. 15
7. 8
9. 7
11. 21 used computers
13. 15 surveys
15. 5.1 miles per hour

2.7

1. 40 ft
3. 415 in.^{2}
5. SA = 6079 in^2; V = 44,580 in^3
7. 30 m
9. 28.26 ft^2
11. 30 degrees
13. 18 bags
15. 22 seconds

2.8

1. $>$
3. $>$
5. 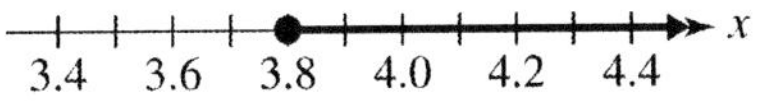

7. $x < -14$
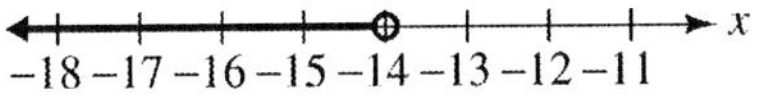

9. $x > 13$
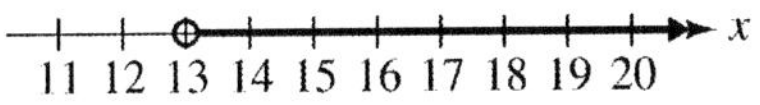

11. $x < 12$
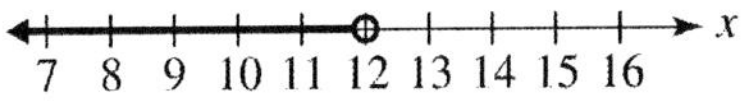

13. $y \geq -\frac{2}{5}$
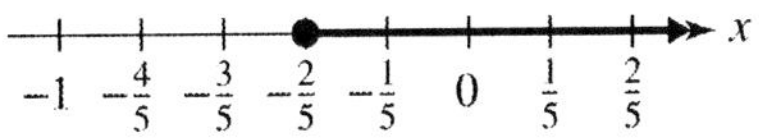

15. $x > -9$

Practice Set Answers Chapter 3

3.1

1., 3., 5.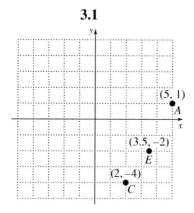

7. (−2, −4)
9. (3, −1)
11. a. −2
    b. 4
13. a. −4
    b. −7
15. a. −8
    b. 2

3.2

1.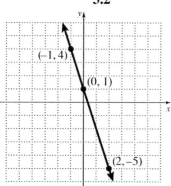

3.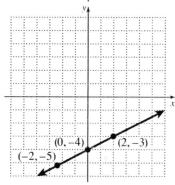

5.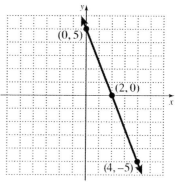

7.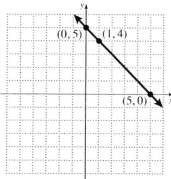

9.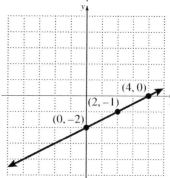

11.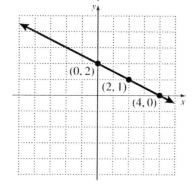

3.3

1. 1
3. $-\dfrac{1}{3}$
5. $\dfrac{3}{4}$, $(0, -4)$
7. $-\dfrac{1}{3}$, $(0, 2)$
9. $y = -4x - \dfrac{4}{5}$
11. $y = \dfrac{6}{5}x + 4$
13.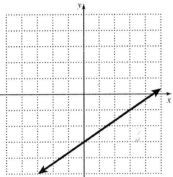
15. a. 3
    b. $-\dfrac{1}{3}$

3.4

1. $y = 3x + 5$
3. $y = -x + 4$
5. $y = -\dfrac{1}{2}x + \dfrac{5}{2}$
7. $y = -\dfrac{1}{2}x + \dfrac{3}{2}$
9. $y = 5x - 11$
11. $y = \dfrac{2}{5}x + \dfrac{6}{5}$
13. $y = -3x - 2$
15. $y = -5x + 8$

3.5

1.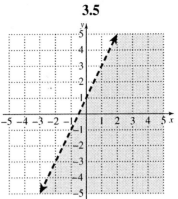
3.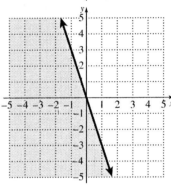
5.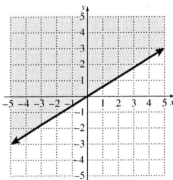
7.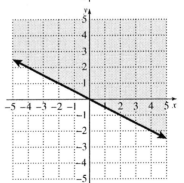

9.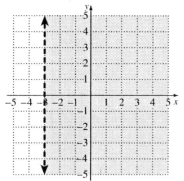

11.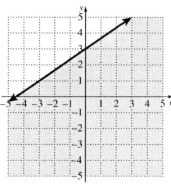

3.6

1. Domain: $\{2, \frac{1}{2}, -4, 5\}$, Range: $\{3, 4, -3, -7\}$; function
3. Domain: $\{2.5, 3.5, 5.5, 8.5\}$, Range: $\{3, 0, -2, -6\}$; function
5.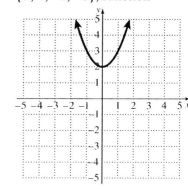

7.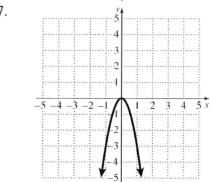

9. not a function
11. a. −2
    b. 13
    c. 4
13. a. 1
    b. 28
    c. 10

Practice Set Answers Chapter 4

4.1

1.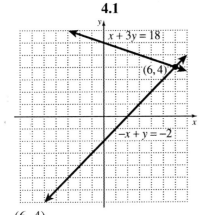
   (6, 4)

3.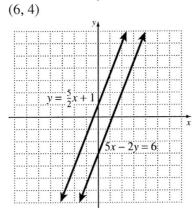
   No solution

5. (−1, 2)
7. No solution; inconsistent system of equations
9. (2, 10)
11. Infinite number of solutions; dependent equations

4.2

1. (0, 1, 6)
3. (1, 4, 0)
5. (2, 7, 1)
7. (−1, −1, 1)
9. $\left(3, \dfrac{2}{3}, -1\right)$
11. (4, −2, 4)
13. (6, 6, 1)
15. Infinite number of solutions; dependent equations

4.3

1. 71, 31
3. $32 per hour for the carpenter, $12 per hour for the helper
5. 185 acres of soybeans, 315 acres of corn
7. Bagel: $0.90; coffee: $0.65
9. Speed of wind: 30 mph; speed of plane: 210 mph
11. 7 free throws, 9 2-point shots
13. 122 adults, 80 students, 48 senior citizens

4.4

1.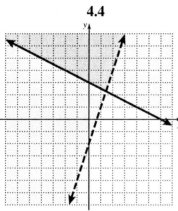

3.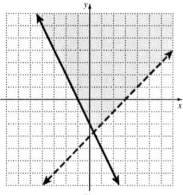

5.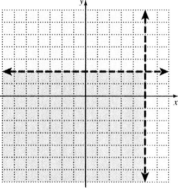

7.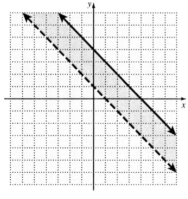

9.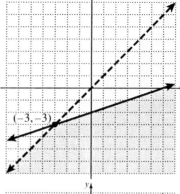

11.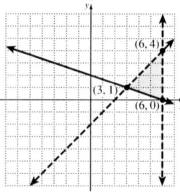

Practice Set Answers Chapter 5

5.1

1. 3^7
3. z^{20}
5. $8a^7b^5$
7. 0
9. x^7
11. $\dfrac{x^6}{-2y^6}$
13. $\dfrac{1}{49x^6}$
15. x^2y^6

5.2

1. $\dfrac{1}{x^5}$
3. 16
5. $\dfrac{18}{xy^5}$
7. $\dfrac{x^{10}}{32y^{20}}$
9. 5.23×10^2
11. 4.7×10^{-2}
13. 0.000137
15. 2×10^{-1}

5.3

1. 5, monomial
3. 8, binomial
5. $-4x^2 + 4x - 7$
7. $10.8x - 14$
9. $3x + 6$
11. $-\dfrac{5}{12}a^2 - \dfrac{7}{30}a + 17$
13. $-2.3x^3 + 9.7x^2 - 4x + 10$
15. 14.56

5.4

1. $-12y^4 + 20y$
3. $-6x^5 + 12x^4 - 21x^3$
5. $10x^7 + 6x^6 - 4x^5 - 2x^4 + 18x^3$
7. $x^2 + 9x + 20$
9. $x^2 - 14x + 33$
11. $12y^2 - 5y - 72$
13. $27x^2 + 3x - 2$
15. $8x^4 - 2x^2y^3 - 15y^6$

5.5

1. $x^2 - 9$
3. $x^2 - 100$
5. $16a^2 - 49$
7. $0.04x^2 - 64$
9. $16x^2 + 24x + 9$
11. $x^3 + 2x^2 - 10x + 25$
13. $x^4 - 5x^3 + 11x^2 - 11x + 4$
15. $x^3 - 4x^2 - 7x + 10$

5.6

1. $4y^3 - 2y^2 + 6$
3. $6x^6 - 2x^4 + 4x$
5. $4y^2 - 6y + 9$
7. $5x + 4$
9. $x^2 - 7x + 2$
11. $x^2 + 2x + 6 - \dfrac{9}{x - 3}$
13. $x^2 + 4x + 4 + \dfrac{3}{2x - 1}$
15. $2x^2 - 5x + 10 - \dfrac{26}{x + 2}$

Practice Set Answers Chapter 6

6.1
1. $3(2x - y)$
3. $4x(x + 1)$
5. $2ab(8 - 6b - 5b^2)$
7. $6x(10xy + 3y - 4)$
9. $9x^2y^2(x + 2)$
11. $(x + 3y)(8a - b)$
13. $(y - 3)(7x + 5)$
15. $(5a - 1)(4x - 3)$

6.2
1. $(y + 4)(y - 2)$
3. $(z + 3)(z - 1)$
5. $(x + 5)(x - 2)$
7. $(x + 5)(y - 2)$
9. $(x - 1)(y - 4z)$
11. $(3x - 2)(2x + 5)$
13. $(5x - 6)(2x + 7)$
15. $(3a + 2b)(4x + 5y)$

6.3
1. $(x + 2)(x + 2)$
3. $(x + 2)(x + 4)$
5. $(x - 6)(x - 2)$
7. $(x - 10)(x - 3)$
9. $(a + 2)(a - 7)$
11. $(a + 1)(a - 5)$
13. $(y^2 + 5)(y^2 + 6)$
15. $3(x + 7)(x - 4)$

6.4
1. $(2x + 7)(x + 1)$
3. $(2x - 1)(2x - 5)$
5. $(3x + 1)(x - 5)$
7. $(2x - 3)(2x + 1)$
9. $(3y + 2)(2y - 1)$
11. $(7x + 1)(x - 2)$
13. $2(3x + 4)(2x - 5)$
15. $3(5y - 3)(y + 4)$

6.5
1. $(4x - 1)(4x + 1)$
3. $(9x - 5)(9x + 5)$
5. $(9y - 1)(9y + 1)$
7. $(3y + 2)^2$
9. $(4y - 3)^2$
11. $4(x - 3)^2$
13. $(x^2 - 10)(x^2 + 10)$
15. $(3x^2 - 5)^2$

6.6
1. $(x + 3)(x + 9)$
3. Prime
5. $(5x - 2y)(5x + 2y)$
7. $2x(x - 5)(x + 9)$
9. $-3xy^2(x + 2)^2$
11. $-x(x + 3)(x - 15)$
13. Prime
15. $(x - 1)(x + 1)(x^2 + 1)$

6.7
1. $-2, 8$
3. $3, -\dfrac{1}{2}$
5. $\dfrac{4}{3}, 0$
7. $0, 3$
9. $7, -5$
11. $\dfrac{1}{3}$
13. $3, 4$
15. $3, -3$

Practice Set Answers Chapter 7

7.1

1. 2
3. $\dfrac{2x+1}{x}$
5. $\dfrac{(2x+3)}{x(x+5)}$
7. $-\dfrac{(y+4)}{(y-2)}$
9. $\dfrac{a-3b}{2a-5b}$

7.2

1. $\dfrac{3(x+1)}{(x-1)}$
3. $\dfrac{x+2}{x-2}$
5. $\dfrac{x}{5x+7}$
7. $\dfrac{x+9}{x+1}$
9. $\dfrac{3(2x-3y)}{4}$

7.3

1. $28(x+3)$
3. $(2x-3)(x-4)(x+5)$
5. $\dfrac{5x-3}{7x-1}$
7. $\dfrac{5x+3}{3x+8}$
9. $\dfrac{7x-2}{3x-3}$
11. $\dfrac{5y+2x}{y(x-y)(x+y)}$
13. $\dfrac{3}{x-7}$

7.4

1. $\dfrac{2x}{3x+1}$
3. $\dfrac{2}{xy}$
5. $\dfrac{2x}{x-35}$
7. $\dfrac{2x-7y}{7}$

7.5

1. 5
3. $-\dfrac{13}{6}$
5. -6
7. no solution
9. -10

7.6

1. 35
3. $23\dfrac{1}{3}$
5. 38
7. $133\dfrac{1}{3}$ mi
9. 16 ft
11. $1\dfrac{1}{2}$ cups
13. 450 km

Practice Set Answers Chapter 8

8.1

1. $\dfrac{4}{9x^4 y^2 z^4}$
3. $z^{9/2}$
5. $a^{7/3}$
7. $y^{1/2}$
9. $\dfrac{1}{7^{1/4}}$
11. $ab^{4/3}$
13. $-3a^{5/3}b - 12a^2$
15. $\dfrac{1 + 7^{1/3} x^{1/2}}{7^{1/3}}$

8.2

1. 9
3. Not a real number
5. 7
7.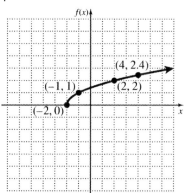
9. $y^{2/3}$
11. $a^{1/12}$
13. $3x^5 y^2$
15. $\dfrac{1}{\sqrt[7]{256}}$

8.3

1. $5\sqrt{2}$
3. $5x\sqrt{x}$
5. 3
7. $-3x^2 yz^2 \sqrt[3]{5y^2 z^2}$
9. $7\sqrt{3}$
11. $64\sqrt{2}$
13. $-4\sqrt{x}$
15. $42 x^2 y^3 \sqrt[3]{x^2}$

8.4

1. $\sqrt{10}$
3. $15a\sqrt{ab}$
5. $8\sqrt{3x} - 14x$
7. $265 + 30\sqrt{70}$
9. $\dfrac{4}{3}$
11. $3x^2 y^2$
13. $\dfrac{2\sqrt{5xy}}{5x}$
15. $\dfrac{2a\sqrt{3} - 2\sqrt{3ab} - \sqrt{2ab} + b\sqrt{2}}{2a - 2b}$

8.5

1. $x = 48$
3. $x = 3$
5. No solution
7. $x = 0$
9. $x = 22$
11. $x = 1$
13. $x = 0, x = 4$
15. $x = 5$

8.6

1. $15i$
3. $3 + i\sqrt{2}$
5. $-6\sqrt{2}$
7. -15
9. $-21 + i$
11. $20 - 3i\sqrt{2}$
13. $-1 + i$
15. $\dfrac{3 - 4i}{2}$

8.7

1. $y = 22.5$
3. 40 inches
5. 196 feet
7. $y \approx 3.9$
9. 12.5 amperes
11. $y \approx 4.2$
13. $q \approx 19.5$
15. 45 pounds

Practice Set Answers Chapter 9

9.1
1. $x = \pm 4$
3. $x = \pm 3\sqrt{5}$
5. $x = -9, x = -23$
7. $x = \dfrac{-1 \pm \sqrt{15}}{3}$
9. $x = -1, x = -5$
11. $x = 3 \pm i\sqrt{21}$
13. $x = \dfrac{-7 \pm \sqrt{73}}{2}$
15. $x = \dfrac{1 \pm i\sqrt{47}}{8}$

9.2
1. $x = -3 \pm \sqrt{15}$
3. $x = \dfrac{-6 \pm \sqrt{15}}{3}$
5. $x = \dfrac{\pm\sqrt{7}}{2}$
7. $x = \dfrac{-5 \pm \sqrt{47}}{2}$
9. $x = \dfrac{-1 \pm i\sqrt{71}}{2}$
11. $x = \dfrac{3 \pm i\sqrt{31}}{5}$
13. 1 rational root
15. $2x^2 + 7x + 3 = 0$

9.3
1. $x = \pm 2\sqrt{2}, x = \pm\sqrt{5}$
3. $x = \pm 1, x = \pm \dfrac{\sqrt[4]{216}}{6}$
5. $x = \sqrt[3]{4}, x = -\sqrt[3]{2}$
7. $x = \sqrt[3]{6}, x = -\sqrt[3]{2}$
9. $x = -1, x = -32$
11. $x = \dfrac{13}{4}, x = \dfrac{5}{4}$
13. $x = 81$
15. $x = \dfrac{5}{2}$

9.4
1. $s = \pm\sqrt{\dfrac{A}{6}}$
3. $r = \pm\sqrt{\dfrac{3V}{\pi h}}$
5. $W = \pm\sqrt{\dfrac{3Jp}{2mn}}$
7. $x = \dfrac{-y \pm \sqrt{y^2 - 160}}{16}$
9. $b = 3$
11. $\dfrac{15\sqrt{2}}{2}$ meters
13. Altitude = 14 inches; base = 5 inches
15. 25 mph on dirt; 45 mph on paved

9.5

1. $V(-3, -1)$; $I(0, 8)$; $(-2, 0), (-4, 0)$
3. $V(-1, 8)$; $I(0, 10)$; no x-intercepts
5. $V\left(-\dfrac{17}{10}, -\dfrac{529}{20}\right)$; $I(0, -12)$; $\left(\dfrac{3}{5}, 0\right)$, $(-4, 0)$

7.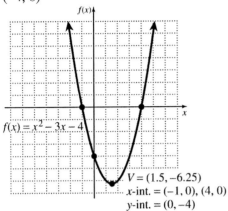
$f(x) = x^2 - 3x - 4$
$V = (1.5, -6.25)$
x-int. $= (-1, 0), (4, 0)$
y-int. $= (0, -4)$

9.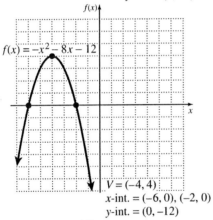
$f(x) = -x^2 - 8x - 12$
$V = (-4, 4)$
x-int. $= (-6, 0), (-2, 0)$
y-int. $= (0, -12)$

11.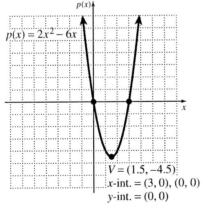
$p(x) = 2x^2 - 6x$
$V = (1.5, -4.5)$
x-int. $= (3, 0), (0, 0)$
y-int. $= (0, 0)$

13.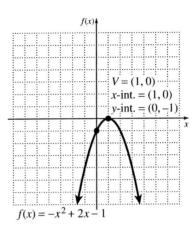
$V = (1, 0)$
x-int. $= (1, 0)$
y-int. $= (0, -1)$
$f(x) = -x^2 + 2x - 1$

9.6

1.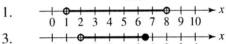
3. (number line from -6 to 4, open at -3, closed at 1)
5. (number line from -5 to 5, closed at -2 and 3)
7. $x > 3$ and $x < 6$
9. $2 < x < 8$
11. $-1 \le x \le \dfrac{5}{2}$
13. $-\dfrac{3}{2} < x < \dfrac{2}{5}$
15. Approximately $x < -6.4$ or $x > 1.4$

9.7

1. $y = 23, y = -23$
3. $x = 0.205, x = -0.205$
5. $x = 0, x = -12$
7. $x = 16, x = 0$
9. $-4 < x < 4$
(number line from -5 to 5, open at -4 and 4)
11. $5 < x < 13$
13. $-8 < x < 12$
15. $x \le -\dfrac{11}{3}$ or $x \ge \dfrac{13}{3}$

Practice Set Answers Chapter 10

10.1

1. 3
3. $3\sqrt{5}$
5. $\dfrac{5}{3}$
7. $(x-3.2)^2 + (y+1.4)^2 = \dfrac{9}{49}$
9.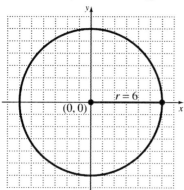
11. $(x+1)^2 + (y-2)^2 = 25$
    center $(-1, 2)$, $r = 5$
13. $x^2 + (y+1)^2 = 4$
    center $(0, -1)$, $r = 2$
15. $\left(x+\dfrac{5}{2}\right)^2 + \left(y+\dfrac{3}{2}\right)^2 = \dfrac{17}{2}$
    center $\left(-\dfrac{5}{2}, -\dfrac{3}{2}\right)$, $r = \dfrac{\sqrt{34}}{2}$

10.2

1.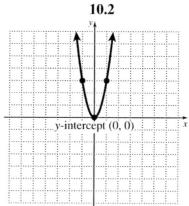
3.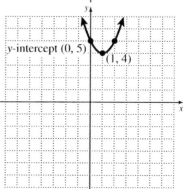
5.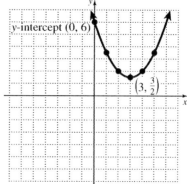
7.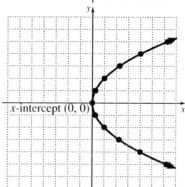

9.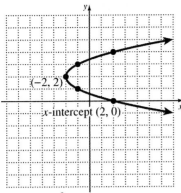

11. $y = (x-4)^2 - 6$
    a. vertical
    b. opens upward
    c. $(4, -6)$

10.3

1.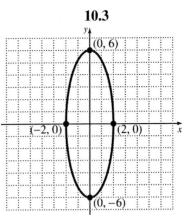

3.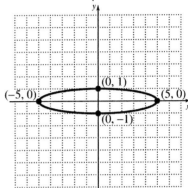

5.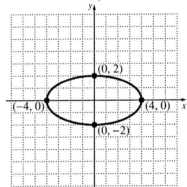

7.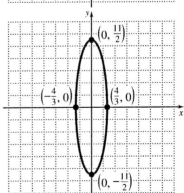

9.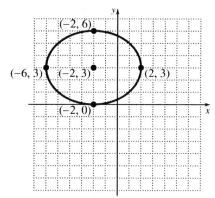

11.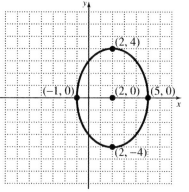

10.4

1.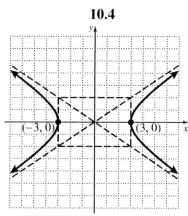

3.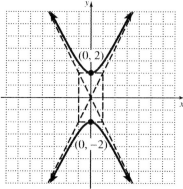

5.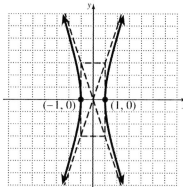

7.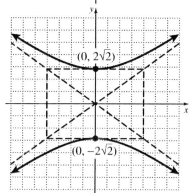

9.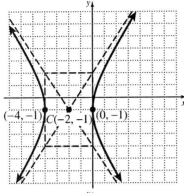

11.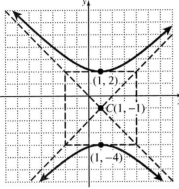

10.5

1. (–2, 0), (3, 5)

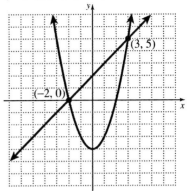

3. (–6, –5), (–5, –6)
5. (6, 0), (5, 1)
7. (3, 5), (–3, –5)
9. (0, 2), (0, –2)

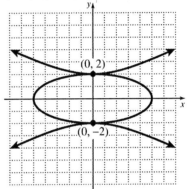

11. (6, 0), (–6, 0)
13. $\left(\dfrac{2\sqrt{15}}{3}, \dfrac{\sqrt{17}}{3}\right), \left(\dfrac{2\sqrt{15}}{3}, -\dfrac{\sqrt{17}}{3}\right),$
$\left(-\dfrac{2\sqrt{15}}{3}, \dfrac{\sqrt{17}}{3}\right), \left(-\dfrac{2\sqrt{15}}{3}, -\dfrac{\sqrt{17}}{3}\right)$

Practice Set Answers Chapter 11

11.1

1. $-\dfrac{13}{3}$
3. $4b^2 - 48$
5. $5a^2 + 2a + 6$
7. $2\sqrt{3}$
9. $\sqrt{b^2 + 3b}$
11. $\dfrac{5}{4a^3 + 3}$
13. 6
15. a. $S(a) = 6a^2$
    b. 96 square feet
    c. $96 + 48e + 6e^2$
    d. 110.94 square feet; the difference in surface area is 14.94 square feet

11.2

1. not a function
3. function
5.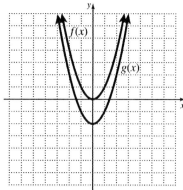
7.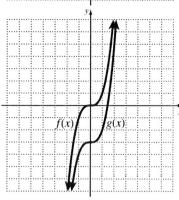
9.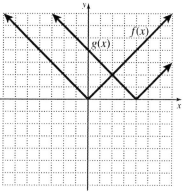
11.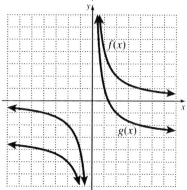

11.3

1. a. $-x + 5$
   b. $7x - 9$
   c. 2
   d. -23
3. a. $1.6x^3 + 4.6x^2 - 2.7x + 7.6$
   b. $1.6x^3 - 4.6x^2 - 2.7x - 7.6$
   c. 84.1
   d. -33.4
5. a. $2x^2 - 6x$
   b. $\dfrac{x - 3}{2x}$
   c. 0
   d. 1
7. a. $x^3 - 12x^2 + 48x - 64$
   b. $x - 4$
   c. -1
   d. -7
9. $1 - 2x$
11. $\dfrac{4}{10x - 9}$

11.4

1. one-to-one
3. one-to-one
5. one-to-one
7. $f^{-1}(x) = \dfrac{x-2}{4}$
9. $f^{-1}(x) = \dfrac{4}{x}$
11. $f^{-1}(x) = \dfrac{x-2}{2}$

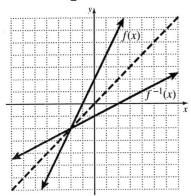

13. $h^{-1}(x) = 2x - 6$

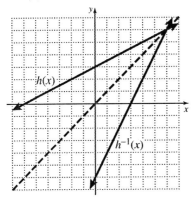

Practice Set Answers Chapter 12

12.1

1.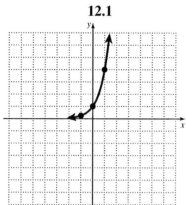

3.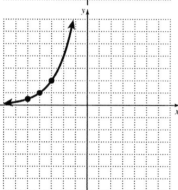

5.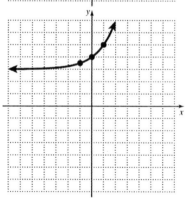

7. $x = 2$
9. $x = 0$
11. $x = -1$
13. $5008.62

12.2

1. $\log_6 216 = 3$
3. $\log_4\left(\dfrac{1}{64}\right) = -3$
5. $49^{1/2} = 7$
7. $x = \dfrac{1}{64}$
9. $x = -4$
11. $x = \dfrac{1}{3}$
13. $\dfrac{1}{2}$
15.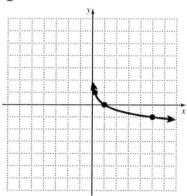

12.3

1. $\log_3 x + \log_3 y$
3. $3 \log_5 y$
5. $\log_2 5 + 3 \log_2 y + \log_2 z - \dfrac{1}{2}\log_2 x$
7. $\log_2 60x$
9. $\log_a\left(\dfrac{36\sqrt[3]{6}}{x^5}\right)$
11. $\dfrac{1}{5}$
13. $x = 11$
15. $x = 30$

12.4
1. 1.049218023
3. 2.106570209
5. 1413.748547
7. 23.44228815
9. 24.5325302
11. 7430.191379
13. 2.432959407
15. −2.456336474

12.5
1. $x = 2$
3. $x = \dfrac{1}{2}$
5. $x = \dfrac{6}{5}$
7. $x = 3$
9. $x = 5$
11. $x = \dfrac{\log 15 - 5 \log 3}{3 \log 3}$
13. $x \approx 6.838$
15. 10 years